TEXTBOOK OF ELECTRONICS

TEXTBOOK
OF
ELECTRONICS

By

Dr. Ajit Kumar Sharma

Deptt. of Physics

D.A.V. (P.G.) College

Bulandshahr

(U.P.)

DPH

DISCOVERY PUBLISHING HOUSE PVT. LTD.

NEW DELHI-110 002

Published by:
Tilak Wasan
DISCOVERY PUBLISHING HOUSE PVT. LTD.
4383/4B, Ansari Road, Darya Ganj
New Delhi-110 002 (India)
Phone : +91-11-23279245, 43596064-65
Fax : +91-11-23253475
E-mail : discoverypublishinghouse@gmail.com
sales@discoverypublishinggroup.com
web : www.discoverypublishinggroup.com

***First Edition:* 2011**

***Reprinted:* 2017**

ISBN: 978-81-8356-836-4

Textbook of Electronics

Printed at:
Infinity Imaging Systems
Delhi

Preface

The book "Text Book of Electronics" has been written the requirements of graduate students of all Indian Universities. It covers the complete syllabus of physics prescribed by Technical Universities. It covers the complete syllabus of electronics prescribed by Technical Universities. I have spared no pains in applying my long experience of degree class teaching to make the book useful to the students.

The subject matter of this book has been presented in this book has sufficient theory and maximum solved examples. One of salient course syllabi features of the present edition of the book is the inclusion of a large number of typical worked out problems which will elucidate various abstract principles and theories discussed in the text.

The author in general hope. That the present book will be warmly received by the students and teachers. We shall indeed be very thankful to our colleague for their recommendations this book of for their students.

—Author

Contents

Transfer of Energy in Electric Circuit, Primary Cells and Secondary Cells, Primary Cells, There are Two Main Defects in the Type of Cells Local Action and Polarisation, A.C. Circuit, Generation of Alternating Voltage and Current, R.M.S. (Root Mean Square Value), Average Value, Form Factor, Amplitude Factor or Crest Factor, Application of A.C. for Basic Elements, AC Series Circuits, Resonance inn R-L-C Circuit, A.C. Parallel Circuit, A.C. Series Parallel Circuit, Generation of Three Phase Voltage, Reasons for Using Three Phase Supply, Star Connection, Delta Connection, Solved Examples.

1

Electromagnetic Field

ELECTRIC DIPOLE MOMENT

It is a vector p whose magnitude is 2aq and direction is from the negative to positive charge.

When the dipole is placed is a uniform external electric field E_1 as shown in Fig. 1 the two charges experience equal and opposite forces

$$\vec{F} = q\vec{E}$$

Thus the net force an the dipole is zero but there is a net torque τ about an axis at point p at right angles to the plane of the paper

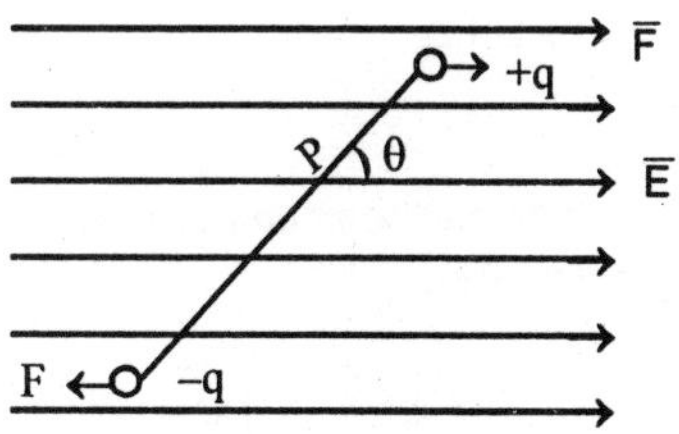

Fig. 1

$$\tau = 2F\,(a \sin \theta)$$
$$= 2(qE)\,(a \sin \theta)$$
$$= 2aqE \sin \theta$$
$$= pE \sin \theta$$

Thus torque is the product of magnitude of force and the perpendicular distance between the forces.

In vector form,

$$\vec{\tau} = \vec{p} \times \vec{E}$$

In order to change the orientation of the electric dipole placed in an external field, some amount of work is to be done and this work is shorted as potential energy in the system.

If the dipole is to be rotated from its reference vertical position. (i.e. θ = 90°) to angle θ, the potential energy is given by.

$$U = \int_{d\theta}^{\theta} p, E \sin\theta \, d\theta = -pE\cos\theta$$

or $U = -\vec{P}.\vec{E}$ ·

THE GAUSS LAW OF ELECTRICITY

The surface integral of the normal component of electric field intensity $\overline{E}$ over a closed surface containing point charge q as shown in Fig. 2 is given by

$$\int \overline{E}.d\overline{S} = q/\epsilon_0$$

This can be interpreted as the net flux of electric field emanating from the surface & containing a point charge q is equal to q/ε_0. It this arbitrary surface does not enclose the point charge the net electric field flux emanating from the surface must be zero i.e.

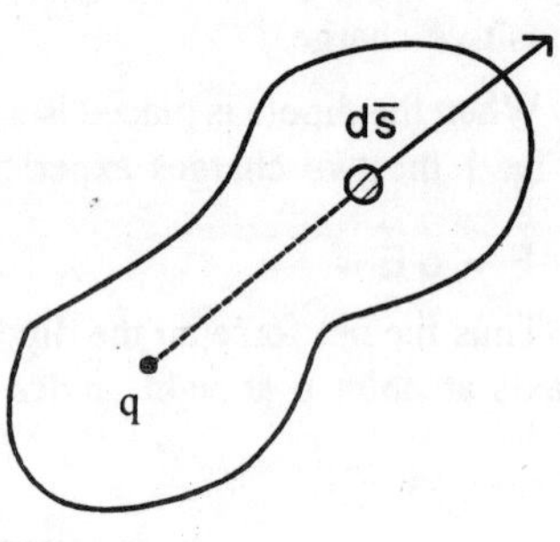

Fig. 2

$$\oint \overline{E}.d\overline{S} = 0$$

If there are more than one point charges enclosed, then the above equation can be generated as follows.

$$\oint_s \overline{E}.d\overline{S} = \oint_S \overline{E}_1.d\overline{S} + \oint_s \overline{E}_2.d\overline{S} + \int_s \overline{E}_2.d\overline{S} + \ldots$$

$$= q_1 + q_2 + q_3 + \ldots q_n = \text{charge enclosed by the surfaces}$$

An important out come of this law is that excess charge placed an insulated conduct for resides entirely an its outer surface.

THE GAUSS'S LAW IN DIFFERENTIAL FROM

Consider a volume distribution with the charge density ρ. The charge enclosed by arbitrary closed surface S is given by the volume integral of charge density through the volume V enclosed by surface, that is $\int_v \rho dv$.

According to Gauss's law.

$$\oint_S \overline{E}.d\overline{S} = \frac{1}{\varepsilon_0}\int_v \rho dv$$

If the volume is shrunk to a very small Δ surface are becomes very small ΔS

$$\lim_{\Delta v\to 0}\oint_{\Delta s}\frac{\overline{E}.d\overline{S}}{\Delta v} = \lim_{\Delta v\to 0}\frac{\left(\frac{1}{\varepsilon_0}\right)\oint \rho dv}{\Delta v}$$

$$= \frac{1}{\varepsilon_0}\lim_{v\to 0}\frac{\rho\Delta V}{\Delta V} = \frac{1}{\varepsilon_0}.\rho$$

or $\Delta\overline{E} = \frac{1}{\varepsilon_0}\rho.$

This equation is Gauss's Law in differential form. It states that the divergence of electric field intensity at any point is equal to $\frac{1}{\varepsilon_0}$ times the volume charge density at that point.

This is Maxwell's divergence equation for the electric field.

ELECTRIC FIELD INTENSITY

The space around the charge which is under stress and experiences a force on another charge when placed there is known as electrostatic field. This is illustrated in Fig. 3(a)

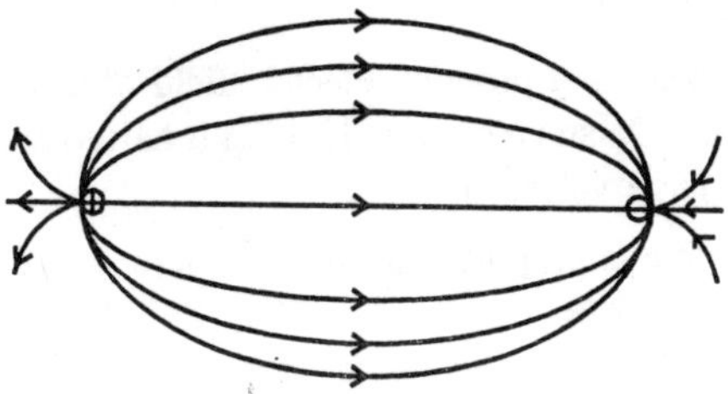

Fig. 3(a)

If the force $\overline{F}$ experienced by a resting positive charge q_0 placed at a point as shown in Fig. 3(b) at a distance R meters from the charge of q, then the field intensity is defined as :

$$E = \frac{\overline{F}}{q_0}$$

The field intensity is a vector in the direction of the force. Its value is given by

$$E = \frac{q}{4\pi \epsilon_0 R^2} \, i_R \; N_w/\text{Coulomb}$$

where i_R is the unit vector along the distance R and directed away from the charge.

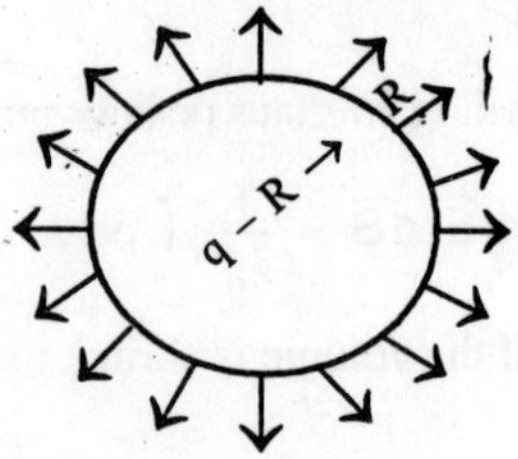

Fig. 3(b)

ELECTRIC DIPOLE

A positive & a negative charge q placed at a distance 2a apart constitutes an electric dipole Fig. 4.

The electric field at a point P due to the dipole will be

$$\vec{E} = \vec{E_1} + \vec{E_2}$$

$$E = 2E_1 \cos\theta$$

$$\& \; E_1 = E_2 = \frac{1}{4\pi \epsilon_0} \cdot \frac{q}{a^2 + r^2}$$

$$\cos\theta = \frac{a}{\sqrt{a^2 + r^2}}$$

$$\therefore \quad E = \frac{1}{4\pi \epsilon_0} \frac{2aq}{\left(a^2 + r^2\right)^{3/2}}$$

if $r \geq a$. $E \cong \dfrac{1}{4\pi \epsilon_0} \dfrac{2a.q}{r^3}$

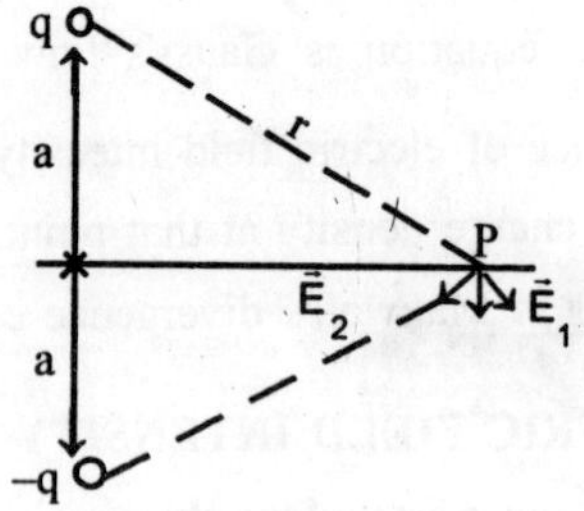

Fig. 4

Since 2a and q appear as a product, electric field of a dipole depends upon this product & not as 2a and q separately. If 2a is halved & q is doubled. E will not change for large values of r.

This E due to a dipole is proportional to $1/r^3$ where as E due to a point charge is proportional to $1/r^3$.

FORM FACTOR KF

It is the ratio of R.M.S. value to the average value of the wave

$$\text{Form factor kf} = \frac{\text{R.M.S. value}}{\text{Average value}}$$

PEAK FACTOR

It is the ratio of peak or maximum value to the R.M.S. value of the wave

$$Kj = \frac{V\,max}{V_{RMS}}$$

The equation of instantaneous alternating current ia and voltage e at any instant t is i = Im sinwt and C = Vm Sinwt where ω is a angular velocity ω = 2 π f radians

THE AVERAGE VALUE OF SINUSOIDAL CURRENT

For the sinusoidal wave form of the current the value at any instant is given by

$$i = Im \sin\theta$$

$$I_{av} = \frac{1}{2\pi}\int_0^{2\pi} I_m \sin\theta$$

$$\text{or} = \frac{I_m}{2\pi}\left[-\cos\theta\right]_0^{2\pi}$$

$$\text{or } I_{av} = \frac{I_m}{2\pi}\left[-1+1\right] = 0$$

Thus average value of sinusoidal wave is zero.

THE RMS VALUE OF ALTERNATING CURRENT

If a steady state alternating current flows through a resistance R for a certain amount of heat is produced. Now assume that direct current is supplied through the same resistance for the same time and same amount of heat is produced this value of steady current is called R.M.S. value of a.c. wave.

$$I_{RMS} = \sqrt{\frac{1}{2\pi}\, i^2\, dt} \qquad = \sqrt{\frac{1}{2\pi}\int_0^{2} I_{m^2} \sin^2 \omega t\, d\omega t}$$

$$= \sqrt{\frac{I_m}{2\pi}\int_0^{2\pi}\left(\frac{1-\cos^2 \omega t}{2}\right) d\omega t}$$

$$= \sqrt{\frac{I_m}{2\pi}\cdot\frac{1}{2}\int_0^{2\pi}\left(\omega t - \frac{\sin^2 \omega t}{2}\right)_0^{2\pi}}$$

$$= \sqrt{\frac{I_m{}^2}{2\pi}\cdot\frac{1}{2}\cdot 2\pi} \qquad = \frac{I_m}{\sqrt{2}}$$

POWER DEFINITIONS AND RELATIONS

Power : The instantaneous power delivered is the product of instantaneous values of applied voltage and current P = Vi

$$\text{Average Power } P = \frac{V\,max}{\sqrt{2}} \cdot \frac{Im\,ax}{\sqrt{2}}$$

P = VI (watts) where V and I are the R.M.S. values

POWER FACTOR

Pf = cos θ , where θ is the phase angle between voltage and current. The power factor can never be greater than unity.

Apparent Power $P_a = V_{RMS}\ I_{RMS}$ (KVA)

True Power $P = P_a \times$ power factor (kw) $= P_a \times \cos\theta$

Reactive power, $P_r = P_a \sin\theta$ (KVAR) where KVAR is kilo - volt - amperes reative

PURELY INDUCTIVE CIRCUIT

When an alternating voltage is applied to purely inductive coil, self induced emf in the coil is induced which opposes the applied voltage. The instantaneous current i is

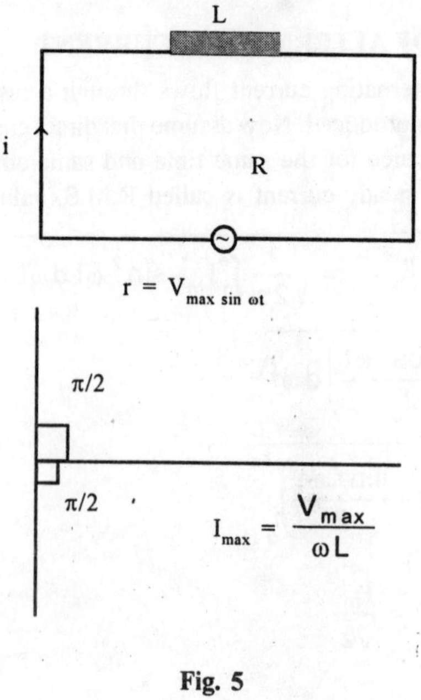

Fig. 5

$$i = I_{max} \sin\left(\omega t - \frac{\pi}{2}\right)$$

It is observed that current lags behind the applied voltage by $\frac{\pi}{2}$ as shown in Fig. 5.

PURELY CAPACITIVE CIRCUIT

When an alternating voltage $v = V_{max}$ sinwt is applied across the capacitor of capacitance C, the instantaneous charge is $q = CV_{max} \sin \omega t$ and instantaneous current is given by

$$i = Im\,ax \sin\left(wt + \frac{\pi}{2}\right)$$

it is observed that current leads the applied voltage by $\pi/2$ and capacitive reactance is $X_C = \frac{1}{\omega_c}$. This is shown in Fig. 6.

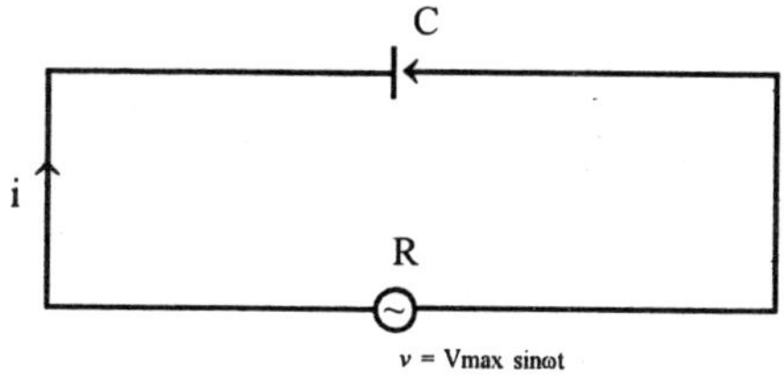

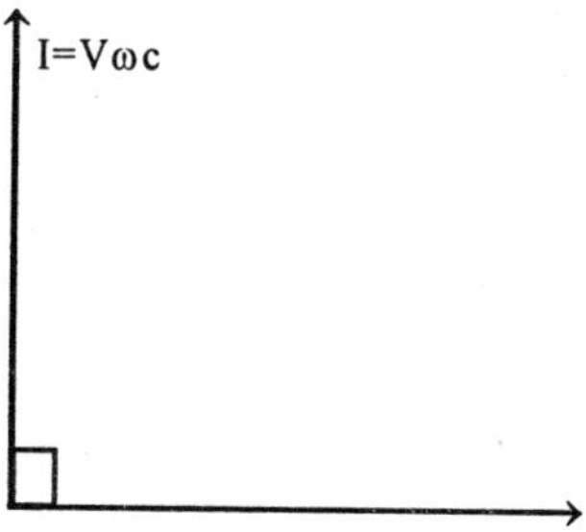

Fig 6

MESH OR DELTA CONNECTION

When the starting end of one coil is connects to the finishing and of another coil, delta connection is obtained. In delta connection, line voltage V_L = phase voltage V_P, and

Line current = $I_L = \sqrt{3}\, I_P$

Total power = $\sqrt{3}\, V_L I_L \cos\theta$

Where θ is the phase angle between respective current and phase voltage

Three wires are carried to the external circuit giving three phase three wires star connected system.

The voltage between any line and neutral point is phase voltage. The voltage between two wires is the line voltage. The neutral point is usually earth connected and the three phases are numbered as R, Y and B.

Potential difference between any two phases = $\sqrt{3}\, V_P$. The current flowing through the lines and phases are same i.e.

$I_L = I_P$

Total Power output = $\sqrt{3}\, V_L I_L \cos\phi$

where ϕ is the angle between respective phase voltage and phase current.

Phase sequence is the order or sequence in which current in different phases attain their maximum values one after the other (The direction of rotation of an induction motor depends on phase sequence, If the phase sequence is reversed by inter changing any two terminals of three phase supply the rotation of the motor would be reversed.

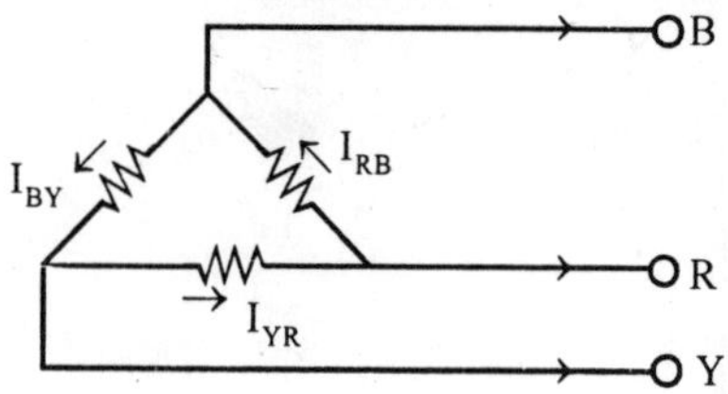

Fig. 7

STAR CONNECTION

When the similar ends are connected together at a common point as shown in Fig. 8 star connection is obtained. The common point N at which similar points are connected together is the neutral point.

Advantages of three phase over single phase are :

(1) The rating of the given machine increases with the increase in number of phase.

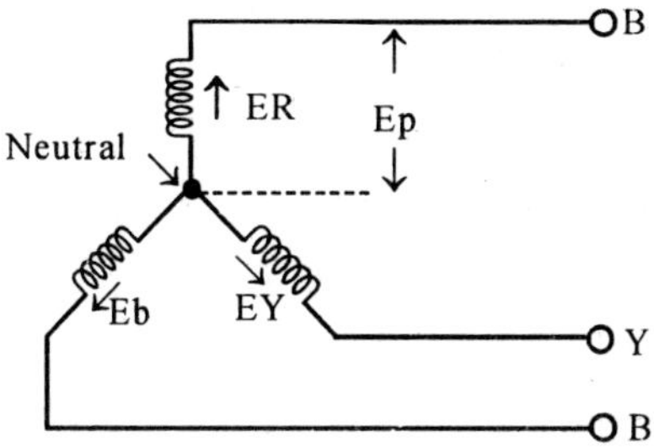

Fig. 8

(2) Single phase motors have no starting torque.

(3) Rotating magnetic field can be set up by passing polyphase current through stationary coils.

(4) Poly Phase system is more capable and reliable then single phase.

(5) The efficiency of poly phase motor is higher than that of single phase motor.

MAGNETOMOTIVE FORCE AND RELUCTANCE

Consider a solenoid of length l meters, N turns cross sectional area A meters^2, wound as a magnetic substance with relative permeability μ_r and carrying a current I amperes.

Field strength inside the solenoid $H = \frac{NI}{l}$ amp-turn per meter

Flux density

$$B = \mu_0\mu_r H = \mu_0\mu_r \frac{NI}{l} \text{ wb/m}^2$$

Total flux created $\phi = B \times a$

$$= \frac{\mu_0\mu_r NI}{l} \times a\text{wb}$$

$$= \frac{NI}{l/\mu_0\mu_r a} \times \text{wb}$$

Now in an electric circuit, current

$$= \frac{\text{emf}}{\text{resis tan ce}}$$

By analogy ampere turn (NI) corresponds to an emf called magneto motive force (mmf) corresponds to resistance known as reluctance i.e. resistance offered to mmf in creating flux in the magnetic material. Reluctance leads to wastage of energy in a magnetic circuit.

$$\text{Therefore, flux} = \frac{\text{mmf}}{\text{reluc tan ce}}$$

Moreover, in an electric circuit, resistance $= \rho \frac{l}{a}$, where ρ = specific resistance of the material conductance of the material is reciprocal of resistance

$$\text{and conductivity} = \frac{1}{\text{specific resis tan ce}}$$

Similarly by analogy permeance of a magnetic material is the reciprocal of its reluctance that is

$$\text{Permeance} = \frac{1}{\text{reluc tan ce}}$$

$$= \frac{\mu_0 \mu_r a}{l} \text{ wb /amp-turn}$$

$$\text{and permeability} = \frac{1}{\text{reluctivit y}}$$

COMPOSITE MAGNETIC CIRCUITS

Consider a circular ring made of different magnetic material of length l_1, l_2 and l_3, cross sectional areas a_1, a_2 and a_3 and relative permeabilities μ_{r_1}, μ_{r2} and μ_{r3} with an air gap of as shown in Fig 9.

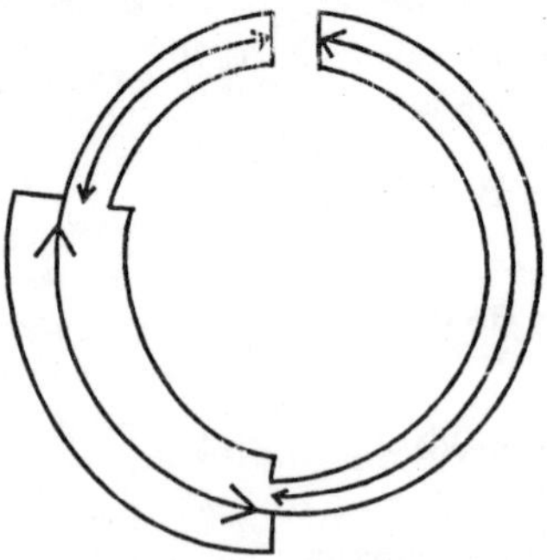

Fig. 9

The total reluctance S of the magnetic circuit will be

$$S = \frac{\ell_1}{\mu_0 \mu_{r_1} a_1} + \frac{\ell_2}{\mu_0 \mu_{r_2} a_2} + \frac{\ell_3}{\mu_0 \mu_{r_3} a_3} + \frac{\ell_g}{\mu_0 a}$$

Total mmf = flux × reluctance

$$= \phi \left[\frac{\ell_1}{\mu_0 \mu_{r_1} a_1} + \frac{\ell_2}{\mu_0 \mu_{r_2} a_2} + \frac{\ell_3}{\mu_0 \mu_{r_3} a_3} + \frac{\ell_g}{\mu_0 a_g} \right]$$

and total Ampere - turns required

$$= H_1 l_1 + H_2 l_2 + H_3 l_3 + H_g l_g$$

ELECTROMAGNET

It is a device consisting of a coil of insulated wire wound an a core usually made of soft iron. The core gets temporarily magnetised when a current flows through the solenoid. As soon as the current is stopped, the device gets demagnetised. Lifting power of an electromagnet is given by

$$P = \frac{B^2 a}{2\mu_0} \text{ Newtons}$$

Where B = Flux density in wb/m^2 and a = area in m^2

Electromagnetic are employed to produce magnetic field is electric generator, small meters, and instruments. It is also used in relays, circuit breakers and motor brakes.

THE COULOMB'S LAW

Charges of same polarity repel one another and that of opposite polarity attract each other. The force (F) between the two charges q_1 & q_2 as shown in Fig. 10 is

(i) Directly proportional to the product of the charges q_1 & q_2.

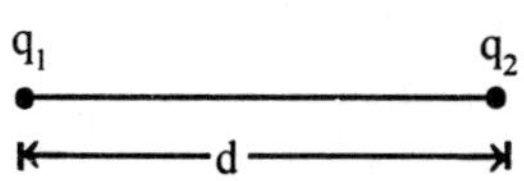

Fig. 10

(ii) Inversely proportional to the square of distance d between them.

(iii) Depends on the nature of medium surrounding the charges mathematically.

$$F \propto \frac{q_1 q_2}{d^2}$$

FLUX OF ELECTRIC FIELD

It refers to a hypothetical surface, closed or open and is measured by the numbers of lines of force cut through the surface. An arbitrary closed surface can be divided into a large number of infinitesimal surfaces represented by a vector $\Delta \overline{S}$ whose direction is at right angles to the small surface as illustrated in Fig. 11. Then its flux can be defined as

$$\Delta\phi_e = \overline{E} . \Delta \overline{S} = E \, \Delta \, S \cos \theta$$

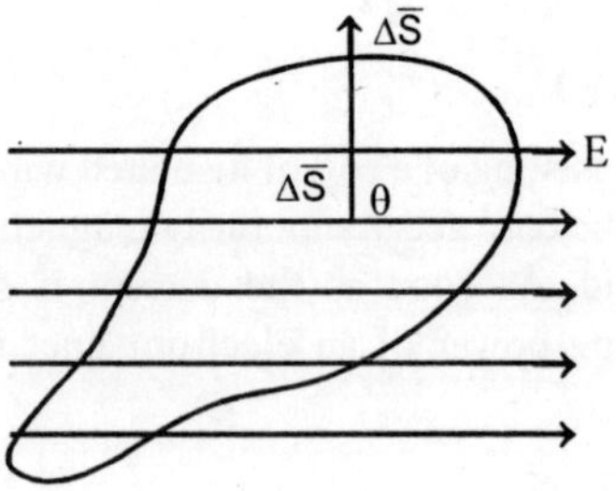

Fig. 11

Which means it is the product of ΔS and component of $\overline{E}$ parallel to vector $\Delta \overline{S}$ or at right angles to the surface. If both vectors point in the same direction. The flux is positive other size negative.

For the whole surface

$$\Delta \phi_e = \int_s \overline{E} . d\overline{s}$$

The electric field intensity of a point charge is thus directed everywhere radically away from the point charge, and an any spherical surface at the point charge, its magnitude is constant.

ELECTRIC FIELD OF MANY CHARGES

If there are several point charges q_1, q_2, q_3 ---- q_n located at different points, then by super position the force $\overline{F}$ experienced by a test charge situated at a point is the vector sum of the forces experienced by the test charge due to the individual charges.

$$\overline{F} \quad \frac{q_1 q}{4\pi \in_0 R_1^0}\, \bar{\iota}_{R_1} + \frac{q_2 q}{4\pi \in_0 R_2^2}\, \bar{\iota}_{R_2} = + \ldots \frac{q_n q}{4\pi \in_0 R_n^2}\, \bar{\iota}_{R_n}$$

The electric field intensity at point P is

$$\overline{E} = \frac{\overline{F}}{q} = \frac{q_1}{4\pi \in_0 R_1^2}\, \bar{\iota}_{R_1} + \frac{q_2}{4\pi \in_0 R_2^2}\, \bar{\iota}_{R_2} + \ldots \frac{q_n}{4\pi \in_0 R_n^2}\, \bar{\iota}_{R_n}$$

$$= \sum_{j=1}^{n} \frac{q_j}{4\pi \in_0 R_j^2}\, \bar{\iota}_{R_j}$$

Z-TRANSFORM

This transform has many properties similar to those of the Laplace transform. The main difference is that the z-transform operates not on the functions of continuous arguments but on the sequence of the discrete integer valued arguments, i.e. $n = 0, \pm 1, \pm 2$, ----.

Sequences that are to be z-transformed must be defined only for $n \geq 0$ and its value for $n < 0$ if any, do not affect the z-transform can be carried further. For every operational rule of Laplace transform, there is a corresponding operational rule of z-transforms and for every application of Laplace transform, there is a corresponding application of z-transform.

Definition : It the function u_n is defined for discrete values (n = 0, 1, 2, ---) and $u_n = 0$ for $n < 0$, then the z-transform is defined to be

$$z\,[u_n] = \overline{u}(z) = \sum_{n=0}^{\infty} u_n\, z^{-n}$$

When ever the infinite series converges.

The inverse z-transform is written as $z^{-1}\,[\overline{u}\,(z)] = u_n$

Example:

Find z-transform of a^n

$$z\left[a^n\right] = \sum_{n=0}^{\infty} a^n\, z^{-n}$$

$$= 1 + (a/z) + (a/z)^2 + (a/z)^3 + \ldots$$

$$= \frac{1}{\left(1 - \left(\frac{a}{z}\right)\right)} = \frac{z}{z - a}$$

$z\,(n^P) = -z\,\dfrac{d}{dz}\, z\,(n^{P-1})$, P being +ve integer.

$$z\left(n^{P}\right)=\sum_{n=0}^{\infty} n^{P} z^{-n}=z \sum_{n=0}^{\infty} n^{P-1} . n . z^{-1(n+1)}$$

Changing P to P – 1, we get

$$z\left(n^{P-1}\right)=\sum_{n=0}^{\infty} n^{P-1} z^{-n}$$

differentiating it w.r.t. z

$$\frac{d}{dz}\left[z\left(n^{P-1}\right)\right]=\sum_{n=0}^{\infty} n^{P-1} .(-n)\, z^{-(n+1)} \qquad ...(2)$$

Substituting z into 1 we get

$$z\left(n^{P}\right) \Rightarrow -z \frac{d}{d_{z}}\left[z\left(n^{P-1}\right)\right]$$

In particular we have the following formula

$$z(1)=\frac{z}{z-1} \text{ (taking a = 1)}$$

$$z(n)=\frac{z}{(z-1)^{2}}$$

$$z\left(n^{2}\right)=\frac{z^{2}+z}{(z-1)^{3}}$$

$$z\left(n^{3}\right)=\frac{z^{3}+4 z^{2}+z}{(z-1)^{4}}$$

$$z\left(n^{4}\right)=\frac{z^{4}+11 z^{3}+11 z^{2}+z}{(z-1)^{5}}$$

Linearly Property

If a, b, c be any constants and u_n, v_n w_n by any discrete functions, then

$z\,(au_n + bv_n - (w_n) = az\,(u_n) + bz\,(v_n) - cz\,(w_n)$

by definition

$z\,(a\,u_n + b\,v_n - c\,w_n)$

$$=\sum_{n=0}^{\infty}\left(a\, u_{n}+b\, v_{n}-c\, w_{n}\right) z^{-n}$$

$$=a \sum_{n=0}^{\infty} u_{n}\, z^{-n}+b \sum_{n=0}^{\infty} u_{n}\, z^{-n}-c \sum_{n=0}^{\infty} w_{n} z^{-n}$$

$= a\ z\ (u_n) + b\ z\ (v_n) - c\ z\ (w_n),$

Damping Rule

If $z\ (u_n) = \overline{u}\ (z)$ then $z\ (a^{-n} u_n) = \overline{u}\ (az)$ by definition $z\ (a^{-n}\ u_n)$

$$= \sum_{n=0}^{\infty} a^{-n}\ u_n z^{-n}$$

$$= \sum_{n=0}^{\infty} u_n\ (az)^{-n}$$

$\Rightarrow \overline{u}\ (az)$

Similarly

$z\ (a^n\ u_n) = \overline{u}\ (z/a)$

$\because$ The geometric factor a^{-n} when $|a|<1$, damps the function u_n, hence damping rule

SOLVED EXAMPLES

Example 1:

Identify the ROC associated with the z-transform for each of the following

$$x[n] = \left(-\frac{1}{2}\right)^n u[-n] + 2\left(\frac{1}{4}\right)^n u[n]$$

$$y[n] = \left(-\frac{1}{2}\right)^n u[n] + 2\left(\frac{1}{4}\right)^n u[n]$$

$$w[n] = \left(-\frac{1}{2}\right)^n u[-n] + 2\left(\frac{1}{4}\right)^n u[-n]$$

Solution:

$$X(z) = \sum_{-\infty}^{0} \left(-\frac{1}{2z}\right)^n + 2\sum_{n=0}^{\infty} \left(\frac{1}{4z}\right)^n$$

$$\Rightarrow \sum_{n=0}^{\infty} (-2z)^k + 2\sum_{n=0}^{\infty} \left(\frac{1}{4z}\right)^n$$

The first series converges for $|z| < \frac{1}{2}$ while second converges for $|z| > \frac{1}{4}$

both series must converge for X(z) to converge, so the ROC is

$\frac{1}{4} < |z| < \frac{1}{2}$ & we set

$$X(z) \Rightarrow \frac{1}{1+2z} + \frac{2z}{z-\frac{1}{4}}$$

which has a pole at $z = \frac{-1}{2} + z = \frac{1}{4}$. The ROC is in the ring shaped region located between the poles.

$$Y(z) \Rightarrow \sum_{n=0}^{\infty} \left(\frac{-1}{2z}\right)^n + 2 \sum_{n=0}^{\infty} \left(\frac{1}{4z}\right)^n$$

First series converges for

$|z| < \frac{1}{2}$ while second $|z| > \frac{1}{4}$ Hence the combined ROC is $|z| < \frac{1}{2}$

$$Y(z) = \frac{z}{z+\frac{1}{2}} + \frac{2z}{z-\frac{1}{4}}$$

for which the poles at $z = \frac{-1}{2}$ & $z = \frac{1}{4}$.

$$W(z) = \sum_{n=-\infty}^{\infty} \left(\frac{-1}{2z}\right)^n + 2 \sum_{n=-\infty}^{0} \left(\frac{1}{4z}\right)^n$$

$$= \sum_{n=\infty}^{\infty} (-2z)^n + 2 \sum_{n=0}^{\infty} (uZ)^n$$

Here the first series converges for

$|2z| < 1$ i.e. $|z| < \frac{1}{2}$

while the second $|z| < \frac{1}{4}$ & the combined is $|z| < \frac{1}{4}$

$$W(z) \Rightarrow \frac{1}{1+2z} + \frac{2}{1-4z}$$

where poles at $z = \frac{-1}{2}$ & $z = \frac{1}{4}$.

The above example shows that ROC of a two sides signal is a ring, the ROC of right sided is extension of a circle & the ROC of left sided signal is the interior of a circle & poles defines the boundaries.

Example 2:

Determine z-transform and ROC of two sided signal $x[n] = z^{|n|}$ assuming $|\alpha| < 1$. Repeat for $|\alpha| > 1$.

Solution:

For $|\alpha| < 1$

$$X(z) = \frac{z}{z-\alpha} - \frac{z}{z-\frac{1}{\alpha}} \quad |\alpha| < |z| < \frac{1}{|\alpha|}$$

for $|a| > 1$, the ROC is empty set.

Properties of z-transform

1. **Linearity :** The linearity properly says that the z-transform of a sum of signals is just the sum of the individual z-transform i.e.

$a\,x[n] + by\,[n] \xleftarrow{z} a\,X(z) + b\,Y(z)$ with ROC is the intersection of individual ROC's because the z-transform of the sum is valid only where both X(z) & Y(z) converges. The ROC can be larger than the intersection if one more terms in x[n] or y[n] cancels each other in the sum. In the z-plane this corresponds to a zero cancelling a pole that defines one of the ROC boundaries.

Example 3:

Suppose $x[n] = \left(\frac{1}{2}\right)^n u[n] - \left(\frac{3}{2}\right)^n u(-n-1)$ Its z-transform is

$$\frac{-z}{\left(z-\frac{1}{2}\right)\left(z-\frac{3}{2}\right)} \text{ with ROC} \frac{1}{2} < |z| < \frac{3}{2}$$

and $y[n] = \left(\frac{1}{4}\right)^n u(n) - \left(\frac{1}{2}\right)^n u[n]$

$$\xleftrightarrow{n} Y(z) = \frac{-\frac{1}{4}z}{\left(z-\frac{1}{4}\right)\left(z-\frac{1}{2}\right)} \text{ with ROC } |z| > \frac{1}{2}.$$

Evaluate z-transform of a x[n] + b y[n]

Solution:

The pole zero plot of X(z) & Y(z) is as shown, Linearity property indicates that

$$a\,x[n] + b\,y[n] \overset{z}{\longleftrightarrow} a\frac{-z}{\left(z-\frac{1}{2}\right)\left(z-\frac{3}{2}\right)} + b\frac{-\frac{1}{4}z}{\left(z-\frac{1}{4}\right)\left(z-\frac{1}{2}\right)}$$

In general the ROC is the intersection of individual ROC's or $\frac{1}{2} < z < \frac{3}{2}$ which corresponds to ROC of X(z) note however, what happens when a = b in this case we have

a x[n] + a y[n]

$$= a\left(-\left(\frac{3}{2}\right)^n u[-n-1] + \left(\frac{1}{4}\right)^n u[n]\right)$$

we see that term $\left(\frac{1}{2}\right)^n u[n]$ has been cancelled in the time domain signal.

The ROC is now easily verified to be $\frac{1}{4} < z < \frac{3}{2}$.

The ROC is larger than the intersection of the individual ROC because the term $\left(\frac{1}{2}\right)^n u[n]$ is no larger present. Combining z-transforms and using linearity properly a X(z) + a Y(z) =

$$a\left[\frac{-z}{\left(z-\frac{1}{2}\right)\left(z-\frac{3}{2}\right)} + \frac{-\frac{1}{4}z}{\left(z-\frac{1}{4}\right)\left(z-\frac{1}{2}\right)}\right]$$

$$\Rightarrow a\frac{-\frac{1}{4}z\left(z-\frac{3}{2}\right) - z\left(z-\frac{1}{4}\right)}{\left(z-\frac{1}{4}\right)\left(z-\frac{1}{2}\right)\left(z-\frac{3}{2}\right)}$$

$$\Rightarrow \frac{-\frac{5}{4}z\left(z-\frac{1}{2}\right)}{\left(z-\frac{1}{4}\right)\left(z-\frac{1}{2}\right)\left(z-\frac{3}{2}\right)}$$

The zero at $z = \frac{1}{2}$ cancels the pole at $z = \frac{1}{2}$ so

$$a\,X(z) + a\,Y(z) = a\frac{-\frac{5}{4}z}{\left(z-\frac{1}{4}\right)\left(z-\frac{3}{2}\right)}$$

Hence cancellation of the $\left(\frac{1}{2}\right)^n u[n]$ term in the time domain corresponds to the cancellation of the pole at $z = \frac{1}{2}$ by a zero in the z-domain. This poled defined the ROC boundary, so the ROC enlarges when it is removed.

Property

Time Reversal $x[-n] \xleftrightarrow{z} X\left(\frac{1}{z}\right)$ with ROC $\frac{1}{R_x}$ Time reversal or reflection, corresponds to replacing z by z^{-1}. Hence if R_x is of the form a < |z| < b, the ROC of the reflected signal is

$a < 1/|z| < b$ or $\frac{1}{b} < z < \frac{1}{a}$.

Time Shift

$x[n - n_0] \xleftrightarrow{z} z^{-n_0} X(z)$

with ROC R_x except possibly z = 0 or |z| = ¥ multiplication by z^{-n_0} introduces a pole of order n_0 at z = 0 if $n_0 > 0$. If is this case the ROC can not include z = 0, even if R_x does include z = 0, unless X(z) has a zero of at least order n_0 at z = 0 that cancels all of the new poles. If $n_0 < 0$, then multiplication by z^{-n_0} introduces n_0 poles at infinity.

If these poles are not cancelled by zeros at infinity in X(z), then ROC of z^{-n_0} X(z) can not include |z| = ¥.

Multiplication by Exponential Sequence

$\alpha^n x[n] \xleftrightarrow{z} X\left(\frac{z}{\alpha}\right)$

Convolution

$x[n] + y[n] \xleftrightarrow{z} X(z)\,Y(z)$

Convolution of time signals corresponds to multiplication of z-transform. As in the linearity property, the ROC may be larger than the intersection if a pole-zero cancellation occurs in the product X(z) Y(z).

Differentiation in the z-domain

$$n\, x[n] \xleftrightarrow{z} -z \frac{d}{dz} X(z)$$

Multiplication by n in time domain corresponds to differentiation with respect to z and multiplication of the result by –z in the z domain. This operation does not charge the ROC.

Example 4:

Find z-transform of signal

$$x[n] = \left(n\left(\frac{-1}{2}\right)^n u[n] \right) \pm \left(\left(\frac{1}{4}\right)^{-n} u[-n] \right).$$

Solution

First we find out the z-transform of

$$w[n] = n\left(\frac{-1}{2}\right)^n u[n]$$

$$\left(\frac{-1}{2}\right)^n u[n] \xleftarrow{z} \frac{z}{z+\frac{1}{2}} \text{ with ROC} > \frac{1}{2}$$

Thus the z domain differentiation properly implies

$$w[n] = n\left(\frac{-1}{2}\right)^n u[n]$$

$$\xleftarrow{z} W(z) = -z\frac{d}{dz}\left(\frac{z}{z+\frac{1}{2}}\right) \text{ with ROC}|z|>1/2$$

$$\Rightarrow -z\left[\frac{z+\frac{1}{2}-z}{\left(z+\frac{1}{2}\right)^2}\right] \Rightarrow \frac{\frac{-1}{2}z}{\left(z+\frac{1}{2}\right)^2} \text{ with ROC } |z| > \frac{1}{2}$$

Next we find the z-transform of $y[n] = \left(\frac{1}{4}\right)^{-n} u[-n]$. We do this by using time reversal property which says $x[-n] \xleftarrow{z} X\left(\frac{1}{2}\right)$ with ROC $\& \frac{1}{R_x}$

$$\left(\frac{1}{u}\right)^n u[+n] \xleftarrow{z} \frac{z}{z - \frac{1}{4}} \text{ with ROC } |z| > \frac{1}{4} \text{ hence}$$

$$y[n] \xleftarrow{z} Y(z) \Rightarrow \frac{z^{-1}}{z^{-1} - \frac{1}{4}} \text{ with ROC } \frac{1}{|z|} > \frac{1}{4}$$

$$\Rightarrow \frac{-4z}{z - z} \text{ with ROC } |z| < 4$$

Lastly we apply the convolution property $x[n] = w[n] + y[n] \xleftarrow{z} X(z)$ $= W(z)\ Y(z)$

$$\Rightarrow \frac{\frac{-1}{2}z}{\left(z + \frac{1}{2}\right)^2} + \frac{-4z}{(z-4)} \frac{1}{2} < |z| < 4$$

$$\Rightarrow \frac{2z^2}{(z-4)\left(z + \frac{1}{2}\right)^2} \frac{1}{2} < |z| < 4$$

Inversion of z-transform

There are two methods for determining inverse z-transform.

1. The method of partial fractions uses knowledge of several basic z-transform pairs and the z-transform properties to invert a large class of z-transform. This approach also relies on an important property of the ROC. i.e. a right sides time signal has a ROC that lies outside the pole radius & left sided time signal has a ROC that lies inside the pole radius.
2. The second inversion method express X(z) as a power series in Z^{-1} of the form

$$X(z) = \sum_{n=-\infty}^{\infty} x[n]\, z^{-n}$$

So that the values of signal may be determined by inspection

$$X(z) = \frac{2z^2 - 2z + 10}{3z^3 - 6z + 9}$$

then we factor z^2 from numerator & $3z^3$ from denominator as

$$X(z) = \frac{z^2}{3z^3}\left(\frac{2 - 2z^{-1} + 10z^{-2}}{1 - 2z^{-2} + 3z^{-3}}\right)$$

$$= \frac{1}{3}z^{-1}\left(\frac{2 - 2z^{-1} + 10z^{-2}}{1 - 2z^{-2} + 3z^{-3}}\right)$$

The partial fraction expansion is of equation (1) is obtained by factoring the denominator polynomial into a product of first orter terms as

$$X(z) = \frac{b_0 + b_1 z^{-1} + \ldots + b_M z^{-M}}{\pi_{k=1}^{N}\left(1 - d_x z^{-1}\right)}$$

Where d_x are the poles of X(z). If none of the poles are repeated, then we may rewrite X(z) as a sum of first-order terms using the partial fractions expansion 1

$$X(z) = \sum_{n=1}^{N} \frac{A_k}{1 - d_k z^{-1}}$$

The inverse z-transform associated with each term is then determined using the transform pair.

Partial Fraction Expansions

In the study of LTI system, we often encounter z-transform that are a rational function of z^{-1}

$$X(z) = \frac{B(z)}{A(z)}$$

$$= = \frac{b_0 + b_1 z^{-1} + \ldots + b_M z^{-M}}{1 + z_1 z^{-1} + \ldots + a_N z^{-N}}$$ and assu-me that $M < N$. If $M \geq N$, the we may use long division to express X(z), in the form

$$X(z) = \sum_{k=0}^{M-N} f_k z^{-k} + \frac{B^2(z)}{A(z)}$$

The numerator polynomial $\tilde{B}$ (z) now has order one less than that of denominator polynomial, and the partial fraction expansion method is applied to determine the inverse transform of $\tilde{B}$ (z)/A(z). The inverse z-transform of the term in the sum are obtained from the pair f(n) $\xleftarrow{z}$ 1 and time shifted property.

In some problems x(z) may be expressed as a ratio of polynomials is z rather than z^{-1}. In this case we may use the partial fraction expansion method described here if we first convert x(z) to a ratio of polynomials is z^{-1}.

$$A_k\,(d_k)^n\,u[n] \xleftarrow{z} \frac{A_k}{1-d_k\,z^{-1}} \text{ with ROC } |z| > d_k$$

or

$$-A_k\,(d_k)^n\,u(-n-1) \xleftarrow{z} \frac{A_k}{1-d_k\,z^{-1}} \text{ with ROC } |z| < d_k$$

The relationship between the ROC associated $|z| < d_x$ with X(z) and each pole determines whether the signal is right sided or left sided inverse transform is chosen for each term.

Condition : If a pole di, is repeated r times, then there are r terms in the partial fraction expansion associated with this pole. They are

$$\frac{Ai_1}{1-di\,z^{-1}}, \frac{Ai_2}{\left(1-di\,z^{-1}\right)^2}, \cdots \frac{Air}{\left(1-di\,z^{-1}\right)^r}$$

The z-transform of each term there is determined using either

$$A\,\frac{(n+1)\ldots(n+m-1)}{(m-1)!}\,(di)^n\,u[n]$$

$$\xleftarrow{z} \frac{A}{(1-di)^m} \text{ with ROC } |z| > di$$

or

$$-A\,\frac{(n+1)\ldots(n+m-1)}{(m-1)!}\,(di)^n\,u(-n-1)$$

$$\xleftarrow{z} \frac{A}{(1-di)^m} \text{ with ROC } |z| < di$$

Again the ROC of X(z) determines whether the right or lift sided inverse transform is chosen.

The linearity property indicates that the ROC of X(z) is the intersection of the ROCs associated with the individual terms in the partial fraction expansion.

In order to choose the correct inverse transform, we must infer the ROC of each term from the ROC of each term from the ROC of X(z). This accomplished by comparing each pole location to the ROC of X(z). If the ROC of X(z) has radius greater than the pole associated with the given terms, we choose right sided inverse transform if the ROC of X(z) has radius less than that of pole we choose left sided inverse transform for this term.

Example 5:

Find the inverse transform of

$$X(z) = \frac{1 - z^{-1} + z^{-2}}{\left(1 - \frac{1}{2}z^{-1}\right)\left(1 - 2z^{-1}\right)\left(1 - z^{-1}\right)}$$ with ROC $|<|z| < 2$.

Solution

Use partial fractions expansion to write

$$X(z) = \frac{A_1}{\left(1 - \frac{1}{\alpha}z^{-1}\right)} + \frac{A_2}{\left(1 - 2z^{-1}\right)} + \frac{A_3}{\left(1 - z^{-1}\right)}$$

$$= \frac{A_1\left(1-2z^{-1}\right)\left(1-z^{-1}\right) + A_2\left(1-\frac{1}{\alpha}z^{-1}\right)\left(1-z^{-1}\right) + A_3\left(1-\frac{1}{\alpha}z^{-1}\right)\left(1-2z^{-1}\right)}{\left(1-\frac{1}{\alpha}z^{-1}\right)\left(1-2z^{-1}\right)\left(1-z^{-1}\right)}$$

$$1 - z^{-1} + z^{-2} \text{ Þ } A_1(1 - z^{-1} - 2z^{-1} + 2z^{-2}) + A_2\left(1 - z^{-1} - \frac{1}{2}z^{-1} + \frac{1}{2}z^{-2}\right) + A_3\left(1 - 2z^{-1} + \frac{1}{2}z^{-1} + z^{-2}\right)$$

$$\Rightarrow A_1(1 - 3z^{-1}) + 2z^{-2}) + A_2\left(1 - \frac{2}{\alpha}z^{-1} + \frac{1}{2}z^{-2}\right) + A_3\left(1 - \frac{5}{\alpha}z^{-1} + z^{-2}\right)$$

Given

$\Rightarrow A_1 + A_2 + A_3 = 1$

$A_1 + A_2 + A_3 = 1$

$+ 3A_1 + \frac{3}{2}A_2 + \frac{5}{2}A_3 = +1$

$2A_1 + \frac{1}{2}A_2 + A_3 = 1$

$A_1 = 1 \; A_2 = 2 \; A_3 = -2$

$$X(z) = \frac{1}{1 - \frac{1}{\alpha} z^{-1}} + \frac{2}{1 - 2z^{-1}} - \frac{2}{1 - z^{-1}}$$

Now we find z-transform of each term using the relation between the pole location & the ROC of X(z).

The ROC has the radius greater than $z = \frac{1}{2}$, so this term is the right sided inverse transform

$$\left(\frac{1}{2}\right)^n u[n] \xleftarrow{z} \frac{1}{1 - \frac{1}{2} z^{-1}}$$

The ROC has the radius less than the pole at z = 2 so this term is left sided signal

$$-2\ (2)^n\ u(-n-1) \xleftarrow{z} \frac{2}{1 - 2z^{-1}}$$

Last ROC has radius greater than the pole at z = 1 ∴ has right sided inverse of z-transform

$$-2u[n] \xleftarrow{z} \frac{2}{1 - 2z^{-1}}$$

$$\text{So } x[n] = \left(\frac{1}{2}\right)^n u[n] - 2\ (2^n)\ u[-n-1] - 2u[n].$$

Example 6:

Find inverse z-transform of

$$X(z) = \frac{z^3 - 10z^2 - 4z + 4}{2z^2 - 2z - 4} \quad \text{with ROC } |z| <|.$$

Solution:

For poles $2[z^2 - z - 2] \Rightarrow 2[z^2 + z - 2z - 2]$

$\Rightarrow 2[(z)(z + 1) - 2(z + 1)]$

$\Rightarrow 2[(z + 1)(z - 2)]$

i.e. having poles at z = – 1 & z = 2

Now convert X(z) into a ratio of polynomials is z^{-1} acc. to equation

$$X(z) = \frac{B(z)}{A(z)} = \frac{b_0 + b_1 z^{-1} + \ldots + b_M z^{-M}}{1 + a_1 z^{-1} + \ldots + a_N z^{-N}}$$

taking z^3 from numerator & $2z^2$ from denominator

$$X(z) = \frac{1}{2}z\frac{(1-10z^{-1} - 4z^{-2} + 4z^{-3})}{(1-z^{-1} - 2z^{-2})}$$

The factor $\frac{1}{2}z$ is easily incorporated later using the time shift properly.

Using division to reduce the order of numerator polynomial

$$-2z^{-2} - z^{-1} + 1 \overline{)\,4z^{-3} - 4z^2 - 10z^{-1} + 1\,} \quad -2z^{-1} + 3$$

$$4z^{-3} + 2z^{-2} - 2z^{-1}$$

$$- \quad - \quad +$$

$$-6z^2 - 8z^{-1} + 1$$

$$-6z^2 - 3z^{-1} + 3$$

$$+ \quad + \quad -$$

$$-5z^{-1} - 2$$

Thus

$$\frac{1 - 10z^{-1} - 4z^{-2} + 4z^{-3}}{1 - z^{-1} - 2z^{-2}} = -2z^{-1} + 3 + \frac{-5z^{-1} - 2}{1 - z^{-1} - 2z^{-2}}$$

$$\Rightarrow -2z^{-1} + 3 + \frac{-5z^{-1} - 2}{(1 + z^{-1})(1 - 2z^{-1})}$$

Using partial fractions

$$\frac{-5z^{-1} - 2}{(1 + z^{-1})(1 - 2z^{-1})} \Rightarrow \frac{A}{(1 + z^{-1})} + \frac{B}{(1 - 2z^{-1})}$$

$$\Rightarrow \frac{1}{1 + z^{-1}} - \frac{3}{1 - 2z^{-1}}$$

& thus define $X(z) = \frac{1}{2}z\,W(z)$

$$W(z) \Rightarrow -2z^{-1} + 3 + \frac{1}{1 + z^{-1}} - \frac{3}{1 - 2z^{-1}} \text{ with ROC } |z| <|$$

The ROC has a smaller radius then either pole & so the inverse z-transform of W(z) is

$$W[n] = -2f[n-1] + 3f[n] - (-1)^n u(-n-1) + 3(2)^n u[-n-1]$$

The time shift property indicates

$$x[n] = \frac{1}{2} w[n+1]$$

Thus

$$x[n] = -\infty[n] + \frac{3}{2} f[n+1] - \frac{1}{2}(-1)^{n+1} u[-n-2] + \frac{3}{2}(2)^{n+1} u[-n-2]$$

Power Series Expansion

Here we seek to express X(z) as a power series is z^{-1} or z of the form

$$X(z) = \sum_{n=-\infty}^{\infty} x[n]\, z^{n-n}$$

The value of the signal x[n] is then given by the coefficients associated with z^{-n}. This inversion method is limited to signals that are on sided; that is, with ROCs of the form $|z| < a$ or $|z| > a$. If the ROC is $|z| > a$, then we express X(z) as a power series is z^{-1} so that we obtain a right sided signal. If ROC is $|z| < a$, then we express X(z) as a power series is z and obtain a left sided inverse transform.

An advantage of the power series approach is the ability to find z. inverse z-transforms for signals that are not a ratio of polynomials is z.

Example 7:

Find inverse z-transform of

$$X(z) = \frac{2 + z^{-1}}{1 - \frac{1}{2}z^{-1}} \text{ with ROC } |z| > \frac{1}{2}$$

using a power series expansion.

Solution

Use long division to write X(z) as a power series in z^{-1}, since the ROC indicates that x[n] is right sided we have

$$1 - \frac{1}{2}z^{-1} \; \frac{2 + 2z^{-1} + z^{-2} + \frac{1}{2}z^{-2} + \ldots}{\sqrt{2 + z^{-1}}}$$

$$\frac{\begin{matrix} 2 - z^{-1} \\ - \; + \end{matrix}}{2z^{-1}}, \quad \frac{2z^{-1} - z^{-2}}{z^{-2}}, \quad \frac{z^{-2} - \frac{1}{2}z^{-3}}{\frac{1}{\alpha}z^{-3}}$$

i.e. $X(z) = 2 + 2z^{-1} + z^{-2} + \frac{1}{2}z^{-3} + ...$

thus we conclude

$x[n] = = 0 \; n < 0$

$x[0] = 2, x[1] = 2$

$x[2] = 1, x[3] = \frac{1}{2} \; ...$

If the ROC is changed to $|z| < /12$, less we expand X(z) as power series is z.

$$\frac{-1}{\alpha} z^{-1} + 1 \quad \frac{-2 - 8z - 16z^2 - 32z^3 + ...}{z^{-1} + 2}$$

$$\frac{z^{-1} - 2}{4}, \frac{4 - 8z}{8z}$$

$$\frac{8z - 16z^2}{16z^2}$$

$X(z) = -2 - 8z - 16z^2 - 32z^3 + ...$

thus $x[n] = 0 \; n > 0$

$x[0] = -2, x[-1] = -8$

$x[-2] = -16, x[-3] = -32$

Example 8:

Find this inverse z-transform of

$X(z) = e^{z^2}$ with ROC all z except $|z| = \infty$.

Solution

Using power series expansion

$$e^a = \sum_{n=0}^{\infty} \frac{a^k}{k!}$$

$$\text{we write } X(z) = \sum_{n=0}^{\infty} \frac{(z^2)^n}{n!} = \sum_{n=0}^{\infty} \frac{z^{2n}}{k!}$$

$$\text{thus } x[n] = \begin{cases} 0 & n > 0 \text{ and } n \text{ add} \\ \frac{1}{\left(\frac{n}{2}\right)!} & \text{otherwise} \end{cases}$$

Stability & Casualty [Causality]

The impulse response of a casual system is zero for n < 0. Therefore, the impulse response of a casual system is determined from the T.F. by using right sided inverse transform. A pole that is inside the unit circle is the z-plane ($|d_x|<|$) contributes an exponentially decaying term to the impulse response, while a pole that is outside the unit circle ($|d_x| > 1$) contributes am exponentially increasing term. Alternatively, if we know system is stable, then impulse response is absolutely summable and the DTFT of the impulse response exists. It follows that, the ROC must include unit circle in the z plane. Hence the relationship between a pole location and the unit circle determines the component of the impulse response associated with the pole. The pole inside the unit circle contributes a right sided decaying exponential term to the impulse response, while a pole outside the unit circle contributes a left sided decaying exponential term to the impulse response.

System that are both stable & casual must have all their poles inside the unit circle. A pole that is inside the unit circle is the z plane contributes a right sided or casual decaying exponential term to the impulse response. We cannot have pole outside the unit circle, since the inverse transform of a pole located outside the circle will contribute other a right-sided increasing exponential term, which is not stable, or a left sided decaying exponential, with is not casual.

Transform Analysis of LTI System

The output y[n] of a LTI system is expressed in terms of the impulse response h[n] & the input x[n] using convolution

$$y[n] = h[n] + x[n]$$

We take z-transform on both side

$$Y(z) = H(z)\ X(z)$$

The z-transform has converted convolution of time sequences into multiplication of transforms. Thus the transfer function offers another description of the o/p-o/p char. of a LTI discrete time system. The T.F. may also the viewed as the ratio of z-transform of the output to that of input i.e.

$$H(z) = \frac{Y(z)}{X(z)}$$

This defines for all values of z for which X(z) is more zone.

The Relationship between T.F. & Difference Equation

The transfer function may be obtained directly from the difference equation description of a system. The difference equation relates o/p y[n] & o/p x[n] as

$$\sum_{n=0}^{N} a_k \; y[n-k] = \sum_{n=0}^{M} b_k \; x[n-k]$$

We know that the transfer f^n. H(z) is an eigen value of the system associated with the eigen function z^n. That is, if $x[n] = z^n$, then the o/p of LTI system is $y[n] = z^n H(z)$.

Substituting $x[n-k] = z^{n-k}$ and $y[n-k] = z^{n-k} H(z)$ into the above relation

$$z^n \sum_{n=0}^{N} a_k \; z^{-k} \; H(z) = z^n \sum_{n=0}^{M} b_k \; z^{-k}$$

$$H(z) = \sum_{n=0}^{M} b_k \; z^{-k}$$

$$H(z) = \frac{\sum_{n=0}^{M} b_k \; z^{-k}}{\sum_{n=0}^{N} a_k \; z^{-k}} \Rightarrow$$

The transfer f^n of a system described by a difference equation is a ratio of polynomials is z^{-1} and is thus termed as rational transfer f^n the coefficient z^{-k} is the numerator is associated with $x[n-k]$ is the difference equation. The coefficient of z^{-k} is the denominator polynomial is the coefficient associated with $y[n-k]$ is the difference equation. We can now find out the transfer function given the difference equation. But also to find as difference equation description for the system given a rational T.F.

The T.F. is pole zero form by factory the numerator & denominator in the above equation

$$\frac{\tilde{b}\pi_{n=1}^{M}\left(1-\left(x\; z^{-1}\right)\right)}{\pi_{n=1}^{N}\left(1-d_x\; z^{-1}\right)}$$

where $\tilde{b} = b_p/a_1$.

The impulse response is the inverse z-transform of the transfer equation. However difference equation does not provide any ROC information; so other system characteristic, such as stability or causality, must be known in order to uniquely determine impulse response.

2

Z-Transform and Transformer

NTRODUCTION

For every operational rule of Laplace transform, there is a corresponding operational rule of z-transforms and for every application of Laplace transform, there is a corresponding application of z-transform. This transform has many properties similar to those of the Laplace transform. The main difference is that the z-transform operates not on the functions of continuous arguments but on the sequence of the discrete integer valued arguments, i.e. n = 0, ± 1, ± 2, ----. Sequences that are to be z-transformed must be defined only for $n \geq 0$ and its value for $n < 0$ if any, do not affect the z-transform can be carried further.

Definition : It the function u_n is defined for discrete values (n = 0, 1, 2, ---) and $u_n = 0$ for $n < 0$, then the z-transform is defined to be

$$z\,[u_n] = \overline{u}(z) = \sum_{n=0}^{\infty} u_n\, z^{-n}$$

When ever the infinite series converges.

The inverse z-transform is written as $z^{-1}\,[\overline{u}\,(z)] = u_n$

Example:

Find z-transform of a^n

$$z\left[a^n\right] = \sum_{n=0}^{\infty} a^n\, z^{-n}$$

$$= 1 + (a/z) + (a/z)^2 + (a/z)^3 + \ldots$$

$$= \frac{1}{\left(1-\left(\frac{a}{z}\right)\right)} = \frac{z}{z-a}$$

$$z(n^P) = -z \frac{d}{dz} z(n^{P-1}), \text{ P being +ve integer.}$$

$$z\left(n^P\right) = \sum_{n=0}^{\infty} n^P z^{-n} = z \sum_{n=0}^{\infty} n^{P-1}.n.z^{-1(n+1)} \quad ...(1)$$

Changing P to P – 1, we get

$$z\left(n^{P-1}\right) = \sum_{n=0}^{\infty} n^{P-1} z^{-n}$$

differentiating it w.r.t. z

$$\frac{d}{dz}\left[z\left(n^{P-1}\right)\right] = \sum_{n=0}^{\infty} n^{P-1}.(-n)\, z^{-(n+1)} \quad ...(2)$$

Substituting z into 1 we get

$$z\left(n^P\right) \Rightarrow -z\frac{d}{d_z}\left[z\left(n^{P-1}\right)\right]$$

In particular we have the following formula

$$z(1) = \frac{z}{z-1} \text{ (taking a = 1)}$$

$$z(n) = \frac{z}{(z-1)^2}$$

$$z\left(n^2\right) = \frac{z^2+z}{(z-1)^3}$$

$$z\left(n^3\right) = \frac{z^3+4z^2+z}{(z-1)^4}$$

$$z\left(n^4\right) = \frac{z^4+11z^3+11z^2+z}{(z-1)^5}$$

Linearly Property

If a, b, c be any constants and u_n, v_n w_n by any discrete functions, then

$z(au_n + bv_n - (w_n) = az(u_n) + bz(v_n) - cz(w_n)$

by definition

$z(a\,u_n + b\,v_n - c\,w_n)$

$$= \sum_{n=0}^{\infty}(a\,u_n + b\,v_n - c\,w_n)\, z^{-n}$$

$$= a \sum_{n=0}^{\infty} u_n \; z^{-n} + b \sum_{n=0}^{\infty} u_n \; z^{-n} - c \sum_{n=0}^{\infty} w_n z^{-n}$$

$= a\ z\ (u_n) + b\ z\ (v_n) - c\ z\ (w_n)$,

Damping Rule

If $z\ (u_n) = \overline{u}\ (z)$ then $z\ (a^{-n} u_n) = \overline{u}\ (az)$ by definition $z\ (a^{-n}\ u_n)$

$$= \sum_{n=0}^{\infty} a^{-n} \; u_n z^{-n}$$

$$= \sum_{n=0}^{\infty} u_n \; (az)^{-n}$$

$\Rightarrow \overline{u}\ (az)$

Similarly

$z\ (a^n\ u_n) = \overline{u}\ (z/a)$

$\because$ The geometric factor a^{-n} when $|a|<1$, damps the function u_n, hence damping rule

WORKING PRINCIPLE

it is static device & works only on A.C. There is no rotating part. it transmit electric power from one circuit to another without changing frequency. Voltage and current can be raised or lowered but power is constant in both sides.

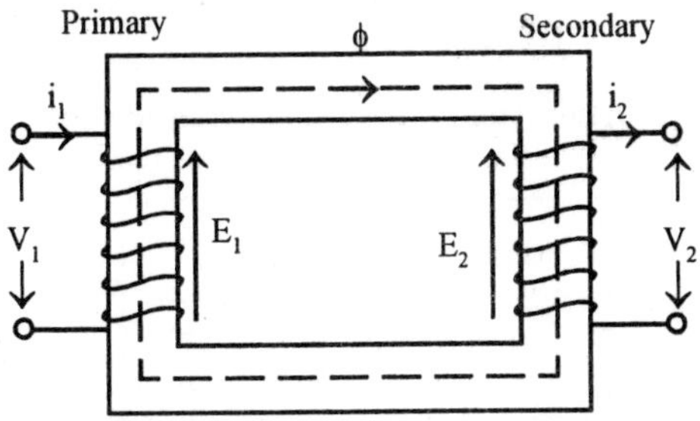

Fig. 1

There are two separate windings called primary winding and secondary winding. These windings are electrically separated but magnetically coupled by core. Thus the flux is common to both the windings. Power from one winding to another is transformed by mutual induction.

When one coil is connected to source of alternating voltage, alternating flux ' ϕ', is set up in the core. This flux is also linked with another coil.

This produces the e.m.f. in another coil, as per Faraday's laws of electromagnetic induction. If the circuit is closed, then current flows and power is transferred from first coil to another.

CONSTRUCTION

Transformers consist of primary winding, secondary winding, core and tank with cooling arrangement.

There are two separate windings made up of number of turns. Both the coils are not electrically connected but they are placed on same core. The coils are also insulated from core.

The core is made up of laminated stripes of transformer steel in order to minimise eddy current loses. Each strip is insulated from each other. High silicon content steel with low hysterisis loop is used in order to minimize hyterisis losses.

The complete assembly is then placed in steel tank. The tank is filled with special insulated oil to cool the transformer. Radiators are provided on the tank to have effective cooling. The transformer coil should be free from moisture. Cooling system may be improved for higher capacity transformer.

Constructionally, there are two types of transformer :

(i) core type

(ii) shell type

(i) **Core type :** In this type, windings surrounds considerable part of core.

(ii) **Shell type :** In this type, core surrounds considerable part of windings.

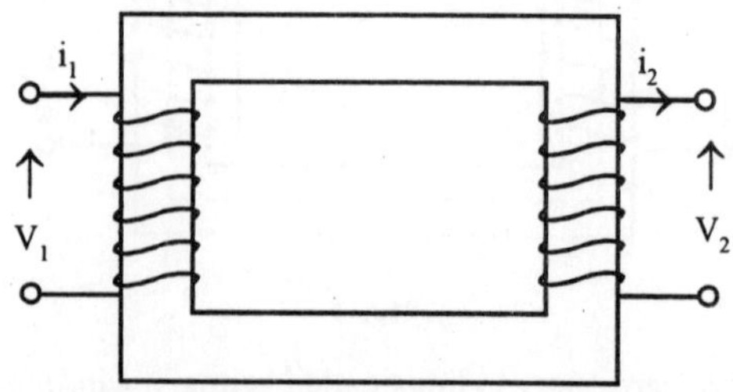

Fig. 2 : Core Type

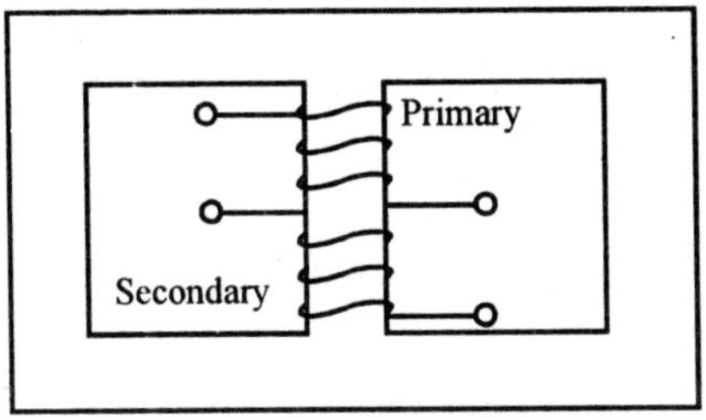

Fig. 3 : Shell Type

E.M.F. EQUATION OF TRANSFORMER (IDEAL TRANSFORMER)

Let us consider ideal transformer without losses and without ohmic resistance. Thus for ideal transformer, there is no voltage & power loss. The magnetic leakage of flux is also absent.

Consider Fig. 3, where

N_1 = No. of turns of primary winding

N_2 = No. of turns of secondary winding

ϕ_m = Maximum flux in core in webers

$\phi_m = B_m \times A$

f = frequency of a.c. input in Hz.

The flux in core increases from zero & reaches to maximum value in $\frac{1}{4f}$ second as shown in Fig. 4.

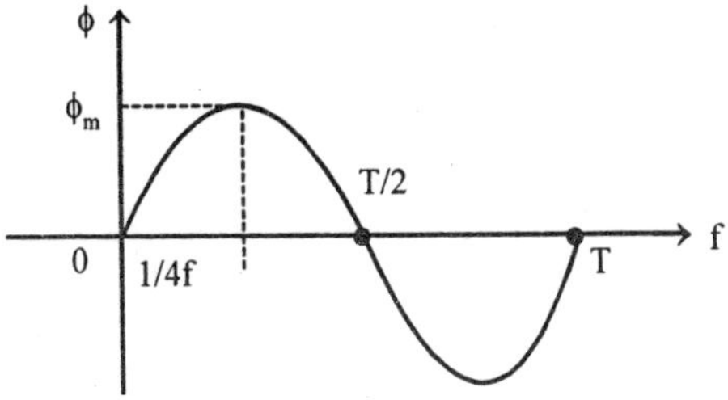

Fig. 4

The average change in flux = ϕ_m

Average change in time = $\frac{1}{4f}$

∴ Average rate of change of flux

$$\Rightarrow \frac{\phi_m}{\frac{1}{4f}} = 4\ f\,\phi_m$$

But rate of change of flux = e.m.f. induced per turn per turn

∴ Average e.m.f. induced/turn

= $4f\phi_m$ volt.

Flux and hence e.m.f. both will varies sinusoidal for sinusoidal wave,

$$\text{Form Factor} = \frac{\text{r.m.s value}}{\text{average value}} = 1.11$$

Thus r.m.s. value of emf/turn

= $1.11 \times 4f\phi_m$

= $4.44\ 4f\phi_m$.

But there are N_1 No. of turns for primary winding.

Thus emf induced in primary (r.m.s.)

= $4.44\ 4fN_1\phi_m$

E_1 = $4.44\ N_1 f\ B_m A$

Similarly emf induced in secondary winding (r.m.s.)

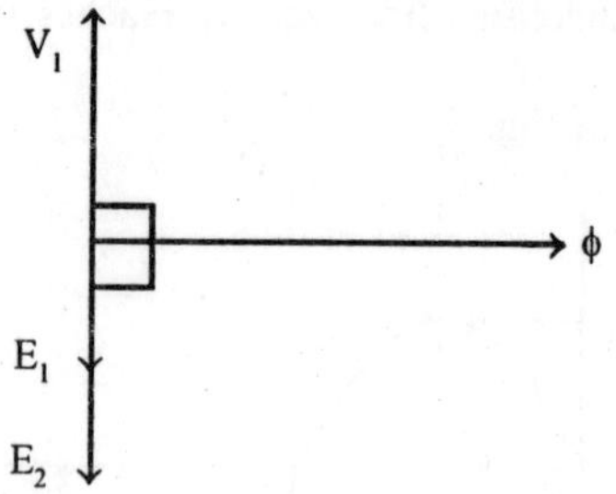

Fig. 5

E_2 = $4.44\ f\ N_2\phi_m$ = $4.44\ f\ N_2 B_m A$

As there is no volatage drop for ideal transformer

$V_1 = E_1$ & $V_2 = E_2$

These voltage are shown in phasor diagram Fig. 5

VOLTAGE & CURRENT TRANSFORMATION RATIO

The ratio of secondary voltage to primary voltage is called voltage transformation ratio (k)

Thus

$$k = \frac{E_2}{E_1} = \frac{4.44 \, f \, N_2 \phi_m}{4.44 \, f \, N_1 \, \phi_m}$$

$$k = \frac{E_2}{E_1} = \frac{N_2}{N_1}$$

But for ideal transformer,

$$E_2 = V_2 \;\&\; E_1 = V_1$$

Thus

$$\frac{E_2}{E_1} = \frac{N_2}{N_1} = \frac{V_2}{V_1} = k$$

Step up transformer : If $N_2 > N_1$, then $k > 1$ or output voltage is more than input voltage & transformer is called step up transformer.

Step down transformer : If $N_2 < N_1$ then $k < 1$ or output voltage is less than input voltage and transformer is called step down transformer.

For ideal transformer, there is is no power loss.

Thus input power = output power

$$V_1 I_1 = V_2 I_2$$

or $$\frac{I_2}{I_1} = \frac{V_1}{V_2} = \frac{1}{k}$$

= current transformation ratio.

TRANSFORMER ON NO-LOAD

In above discussion, we assumed on ideal transformer, where winding resistances, leakage reactances and losses were absent. But this is not the case with actual transformer.

On no load, secondary current is zero. But primary winding carries small current I_0 which is not fully reactive.

I_0 is not 90° behind V_1 but lags it by an angle $\phi_o < 90°$.

Thus no load input power.

$$W_0 = V_1 I_0 \cos \phi_0$$

where V_1 = primary voltage

$\cos \phi_0$ = No load power factor

Thus phasor diagram of transformer on no load is shown in Figure 3.5

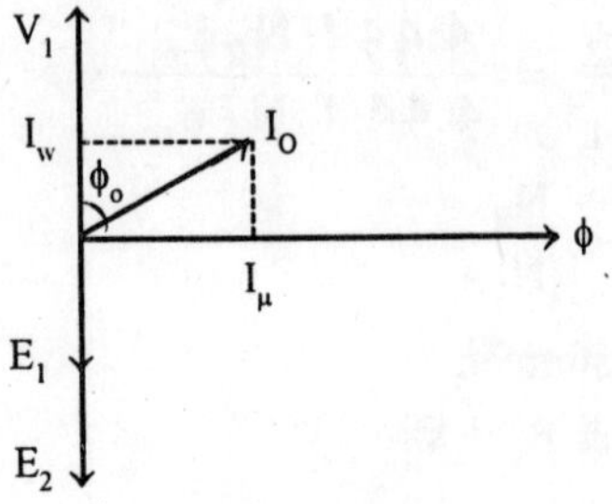

Fig. 6

As shown in Fig. 6, I_0 has two components.

(i) Active or working or Iron loss components $(I_w)_0$ which is in phase with V_1.

Thus $I_w = I_0 \cos \phi_0$

I_w supplies iron losses which consist of eddy current loss and hysterisis loss, small amount of copper losses in primary winding is neglected.

(ii) Idle or reactive or magnetising component (I_μ), which produces the magnetising flux and is in phase with flux

$$I_m = I_0 \sin \phi_0$$

and $I_0 = \sqrt{I_w^2 + I_\mu^2}$.

WINDING RESISTANCES & LEAKAGE REACTANCES

In actual transformer, winding resistance is always present, which causes voltage drop. Primary winding has resistance equal to R_1. Causes voltage drop of I_1R_1. Secondary winding has resistance of R_2, causes voltage drop of I_2R_2.

Similarly all flux linked with primary winding does not link with secondary winding. Some flux completes its path through air & does not link with another coil. These fluxes are called leakage flux. Leakage fluxes increases with load and produces self induced back e.m.f. in respective coils. Thus they are represented by small inductance in series with coils. X_1 is for primary & X_2 for secondary as shown in Fig. 7. These leakage reactances and resistances causes voltage drop in windings.

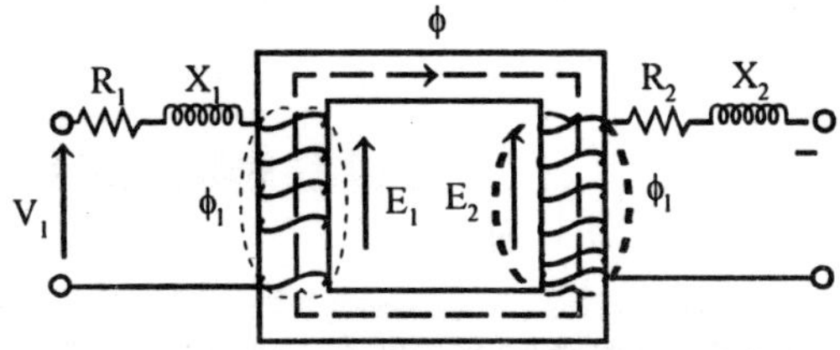

Fig. 7

TRANSFORMER ON LOAD

When transformer is loaded, current I_2 flows in secondary windings. The secondary. Amp turns N_2I_2 set up a secondary flux. This tends to reduce the flux produced by primary Amp turns. As a result of this, the induced e.m.f. in primary E_1 reduces & balance between applied voltage V_1 & E_1 would no longer exist. Hence on load, the presence of secondary amp turns necessitates the production of primary amp turns equal in magnitude but opposite in direction, the above condition is achieved by load current I_2' which flows from supply through primary winding. The primary amp turns with this current is equal to secondary amp turns. Thus e.m.f. are balanced on two sides.

$$I'_2 N_1 = I_2N_2$$

$$\therefore \ I_2' = \left(\frac{N_2}{N_1}\right) I_2 .$$

Hence primary current I_1 drawn from supply is the phasor sum of

(i) No load current I_0 and

(ii) Current I'_2.

EQUIVALENT CIRCUIT OF TRANSFORMER

While drawing equivalent circuit, ideal transformer along with resistances & reactances are considered as shown in Fig. 8. No load current is phasor sum of I_W & I_μ is equivalent to that of current drawn by resistance R_0 & inductive reactance X_0 connected in parallel. The value of R_0 is such that $I_W^2 \ R_0$ is equal to core losses of actual transformer. Similarly value of X_0 is such that current drawn should be equal to I_μ.

As the no load current of transformer is around 5% of full load current, it can be neglected to draw approximate equivalent diagram. The resistances of secondary side can also be transferred to primary side, considering equal power loss.

Let R'_2 is resistance of secondary side referred to primary side.

Then

$$I^2_1 R'_2 = I^2_2 R_2$$

$$R'_2 = \left(\frac{I_2}{I_1}\right)^2 R_2 = \left(\frac{N_1}{N_2}\right)^2 R_2$$

Hence equivalent resistance of transformer referred to primary side.

$$R_{O_1} = R_1 + R'_2$$

$$= R_1 + \left(\frac{N_1}{N_2}\right)^2 R_2$$

Similarly equivalent leakage reactance referred to primary side

$$X_{O_1} = X_1 + \left(\frac{N_1}{N_2}\right)^2 X_s.$$

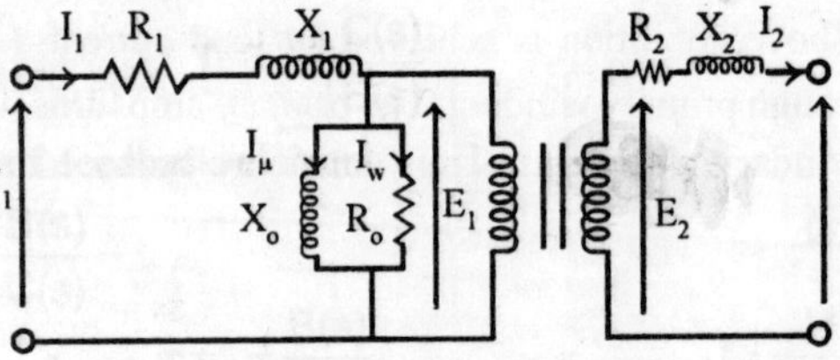

Fig. 8 : Equivalent Circuit

Approximate equivalent circuit is shown in Fig. 3.9

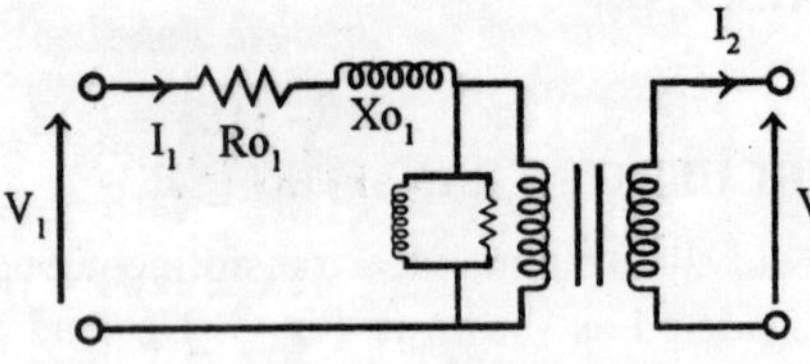

Fig. 9

Similarly equivalent resistances and reactances can be referred to secondary side.

Equivalent resistance referred to secondary side

$$R_{O_2} = R_2 + \left(\frac{N_2}{N_1}\right)^2 R_1$$

$$X_{O_2} = X_2 + \left(\frac{N_2}{N_1}\right)^2 X_1.$$

VOLTAGE REGULATION OF TRANSFORMER

Regulation of transformer is defined as change in secondary voltage between no load & full load condition expressed as a percentage of secondary no load voltage. It is expressed in percentage

$$\text{Regulation} = \frac{\begin{pmatrix}\text{no load voltage}\\ \text{secondary}\end{pmatrix} - \begin{pmatrix}\text{full load voltage}\\ \text{secondary}\end{pmatrix}}{\begin{pmatrix}\text{no load voltage}\\ \text{secondary}\end{pmatrix}}$$

Approximate formula for regulation is

$$R = I_2 \frac{R_{O_2} \cos\theta \pm X_{O_2} \sin\theta}{V_2}$$

Where θ is angle between V_2 & I_2 + ve sign should be taken for lagging power factor & –ve sign should be taken leading power factor.

Regulation should be low for good transformer.

LOSSES IN TRANSFORMER

There are only two type of losses in transformer

(i) **Iron or Core Loss :** This consist of hysterisis & eddy current loss. These are independent of load & hence called constant loss. These can be minimised by using high silicon content steel & lamination.

(ii) **Copper or Winding Loss :** This consists of primary and secondary winding copper loss. These varies with load. Copper loses at half load are equal to one fourth of full load losses.

There are no friction or winding losses.

Iron loss depends on voltage & copper loss depends on current. Hence rating of transformer is given in VA or KVA.

EFFICIENCY OF TRANSFORMER

The ratio of output power to input power is called efficiancy.

$$\%\eta = \frac{\text{Output power}}{\text{Input power}} \times 100$$

As input = output + losses

$$\eta = \frac{\text{Input} - \text{losses}}{\text{Input}} = 1 - \frac{\text{losses}}{\text{Input}}$$

For good transformer, efficiency should be high. Efficiency of transformer is much higher than rotating machines, because there is not friction or windage losses. Efficiency varies with load as shown in Fig. 10.

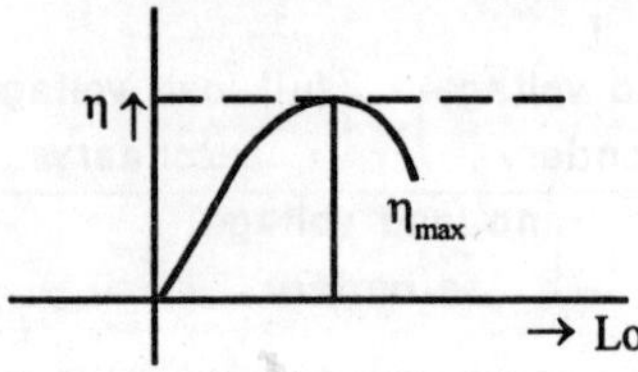

Fig. 10

Maximum efficiency occurs when copper loss = Iron loss.

KVA at

$$\eta_{max} = KVA_{rated} \times \sqrt{\frac{\text{Iron loss}}{\text{Fall load copper loss}}}$$

ALL DAY EFFICIENCY

Distribution transformer are placed in service for 24 hrs.

$$\eta_{all}\ \text{day} = \frac{\text{Output in KWh}}{\text{Input in KWh}} \quad \text{(for 24 hrs)}$$

For a given KVA loading, all day efficiency of transformer is less than commercial efficiency.

OPEN CIRCUIT AND SHORT CIRCUIT TEST

These tests are helpful to determine parameters, efficiency and regulation of transformer at full load or any other load these tests are simple, low power required, accurate and no need of actual loading.

(a) **Open Circuit & Short Circuit Test :**

In this test, secondary is kept open and voltmeter, ammeter & wattmeter are connected in primary side as shown in Fig. 11. Ammeter reads no load primary current I_0. Rated voltage is applied to primary winding. Small copper losses of primary winding is neglected. Thus wattmeter reads constant loss or Iron loss.

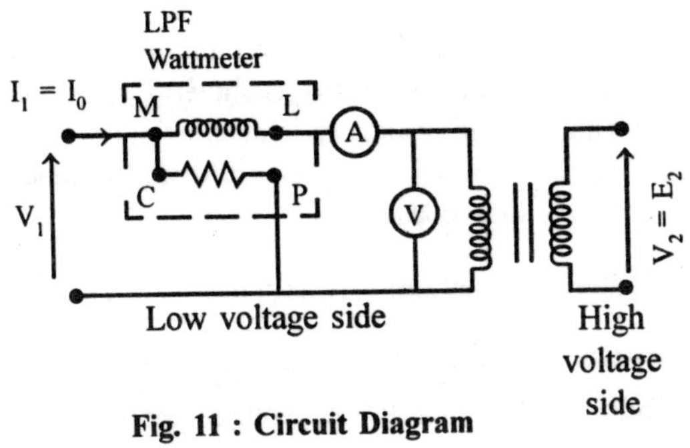

Fig. 11 : Circuit Diagram

Let W_0 = (wattmeter reading) = Iron loss

I_0 = no load current = Ammeter Reading

V_0 = rated supply voltage = voltmeter reading

$W_0 = V_0 I_0 \cos \phi_0$

Thus no load power factor $\frac{W_0}{V_0 I_0} = \cos \phi_0$

Magnetising current $I_\mu = I_0 \sin \phi_0$

Active current component $I_w = I_0 \cos \phi_0$

Magnetising reactance = $X_0 = \frac{V}{I_\mu}$

Resistance equivalent of core loss

$$= R_0 = \frac{V}{I_w}$$

Hence parameter X_0 & R_0 can be determined by o.c. test.

(b) Short Circuit Test :

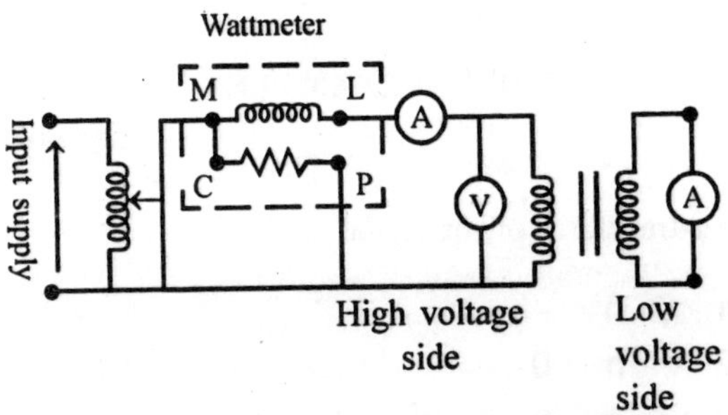

Fig. 12 : Short Circuit Test Circuit Diagram

In this test one of the winding is short circuited by thick wire. Wattmeter, ammeter and voltmeter are connected as shown Fig. 12. Reduced voltage is applied through auto transformer such that ammeter reads rated value. As very less voltage is applied, iron losses are very small & hence neglected. Thus wattmeter gives full load copper loss.

W_{sc} = Wattmeter Reading = $\left(I_{sc}^2\right) R_{O_1}$

I_{sc} = Ammeter Reading = Rated S.C. current

V_{sc} = Voltmeter Reading = S.C. voltage applied.

Equivalent Resistance from primary side $R_{O_1} = \dfrac{W_{sc}}{\left(I_{sc}^2\right)}$

$$Z_{O_1} = \frac{V_{sc}}{I_{sc}}, \quad X_{O_1} = \sqrt{Z_{O_1}^2 - R_{O_1}^2}$$

Knowing equivalent resistance & reactance referred to particular winding, regulation & efficiency at any power factor can be determined.

Auto Transformer or Variac

This type of transformer is used to get variable voltage. This has only one winding. Some part of winding is common to both primary & secondary as shown in Fig. 13. Theory is same as the two winding transformer. This gives saving in copper.

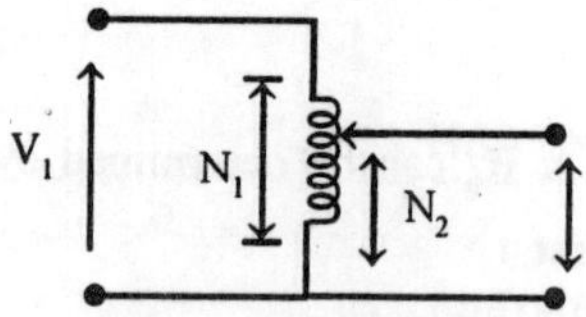

Fig. 13 Auto Transformer

SOLVED EXAMPLES

Example 1:

Determine z-transform of the signal

$$x[n] = \begin{cases} 1 & n = -1 \\ 2 & n = 0 \\ -1 & n = 1 \\ 1 & n = 2 \\ 0 & \text{otherwise} \end{cases}$$

use the z-transform to determine DTFT of x[n].

Solution:

The formula is

$$X(z) = \sum_{n=-\infty}^{\infty} x(n) z^{-n}$$

we get

$X(z) \Rightarrow x(z)\ z^{-2} + x(1)\ z^{-1} + x(0)\ z^{0} + x\ (-1)\ z^{1} + x(-z)\ z^{2}$

$\Rightarrow z + 2 - z^{-1} + z^{-2}$

We obtain the DTFT by putting $z = e^{i\Omega}$ we get

$X(e^{i\Omega}) = e^{i\Omega} + 2 - e^{-i\Omega} + e^{-i2\Omega}$

The most commonly encountered form of the z-transform in engineering applications is a ratio of two polynomials in z^{-1}, as shown by the rational function

$$X(z) = \frac{b_0 + b_1 z^{-1} \ldots + b_M\ z^{-M}}{1 + a_1 z^{-1} + \ldots + a_N\ z^{-N}}$$

It is useful to X(z) as a product of the terms involving the roots of the nòminator and denominator polynomials. i.e.

$$X(z) = \frac{b_0\ \pi_{k=1}^{M} \left(1 - c_k\ z^{-1}\right)}{\pi_{k=1}^{N} \left(1 - d_x\ z^{-1}\right)}$$

The c_k are the roots of the numerator polynomial and are termed the zeros of X(z). The d_x are the roots of denominator polynomials and are termed as poles of X(z). Zero locations are denoted by 'O' symbol and poles locations with the 'x' symbol in the z-plane. The pole and zero locations completely specify X(z) except for the gain factor b_0.

Problem Determine the z-transform of the signal

$x[n] = \alpha^{n}\ u[n]$

Example 2:

Depict the ROC and pole and zero location of X(z) in the z-plane.

Solution:

Substituting x[n] is

$$X(z) = \sum_{n=-\infty}^{\infty} x(n)\ z^{-n}$$

$$= \sum_{n=-\infty}^{\infty} \left(\frac{a}{z}\right)^n$$

This is a geometric series of infinite length in the ratio α/z. The sum converges provided

|α|z|<| or |z|>|α| Hence

$$X(z) = \frac{1}{1 - \alpha z^{-1}} \quad |z| > |\alpha|$$

$$= \frac{z}{z - \alpha}$$

There is a pole at z = α and a zero at z = 0 and the shaded portion in the ROC an z plane.

As with the Laplace transform, the expression for X(z) does not corresponds to a unique time signal unless the ROC is specified. This means that two different signals may have identical z-transform but two different ROC's.

Example 3:

1. $z\left(na^n\right) = \dfrac{az}{(z-a)^2}$

2. $z\left(n^2a^n\right) = \dfrac{az^2 + a^2z}{(z-a)^3}$

3. $z(\cos n\theta) = \dfrac{z(z - \cos\theta)}{z^2 - 2z\cos\theta + 1}$

4. $z(\sin n\theta) = \dfrac{z\sin\theta}{z^2 - 2z\cos\theta + 1}$

5. $z(a^n\cos n\theta) = \dfrac{z(z - a\cos\theta)}{z^2 - 2az\cos\theta + a^2}$

6. $z(a^n\sin n\theta) = \dfrac{az\sin\theta}{z^2 - 2az\cos\theta + a^2}$

Solution:

(1) We know that $z(n) = \dfrac{z}{(z-1)^2}$

$$z(na^n) = \bar{u}\left(a^{-1}z\right) \Rightarrow \frac{a^{-1}z}{\left(a^{-1}z-1\right)^2}$$

$$\Rightarrow \frac{az}{(z-a)^2}$$

(2) $z(n^2) = \dfrac{z^2+z}{(z-1)^3} = \bar{u}(z)$

$$z\,(n^2a^n) = \bar{u}\,(a^{-1}z) \Rightarrow \frac{\left(a^{-1}z\right)^2 + a^{-1}z}{\left(a^{-1}z-1\right)^3}$$

$$= \frac{\left(a^{-1}z\right)^2 + a^{-1}z}{\left(a^{-1}z-1\right)}.$$

We know that $z\,(\cos n\theta)$

$$= \frac{z(z-\cos\theta)}{z^2 - az\cos\theta + 1}$$

$$z(a^n\cos n\theta) = \frac{a^{-1}z\left(a^{-1}z-\cos\theta\right)}{\left(a^{-1}z\right)^2 - z\left(a^{-1}z\right)\cos\theta + 1}$$

$$= \frac{z(z-a\cos\theta)}{z^2 - 2az\cos\theta + a^2}.$$

(1) Shifting u_n to the right

If $z(u_n) = \bar{u}(z)$ then $z(u_{n-k}) = z^{-k}\,\bar{u}(z)$ for $k > 0$

By definition

$$z(u_{n-k}) = \sum_{n=0}^{\infty} u_{n-k}\, z^{-n}$$

$$= u_0\, z^{-k} + u_1 z^{-(k+1)} + \ldots$$

$$= z^{-k}\sum_{n=0}^{\infty} u_n\, z^{-n} = z^{-k}\,\bar{u}(z)$$

This rule is applicable/or very useful to difference equations.

(2) Shifting u_n to the left

of $z(u_n) = \bar{u}(z)$ then

$$z(u_{n+k}) = z^k\,[\overline{u}(z) - u_0 - u_1 z^{-1} - u_2 z^{-2} \ldots - u_{k-1}\, z^{-)k-1)}].$$

Proof

$$z(u_{n+k}) = \sum_{n=0}^{\infty} u_{n+k}\; z^{-n}\;, = z^k \sum_{n=0}^{\infty} u_{n+k}\; z^{-n} = z^{-(n+k)}$$

$$= z^k\left[\sum_{n=0}^{\infty} u_n\; z^{-n} - \sum_{n=0}^{k-1} u_n\; z^{-n}\right]$$

Hence $z(u_{n+k}) = z^k\; \overline{u}\,(z) - u_0 - u_1 z^{-1} - u_2 z^{-2} \ldots. - u_{k-1}\; z^{-(k-1)}]$

The following useful results are

(1) $z(u_{n+1}) = z(\overline{u}\,(z) - u_0)$

(2) $z(u_{n+k}) = z^2(\overline{u}\,(z) - u_0 - u_1 z^{-1})$

(3) $z(u_{n+3}) = z^3(\overline{u}\,(z) - u_0 - u_1 z^{-1} - u_2 z^{-2})$.

Example 4:

Show that $z\left[\dfrac{1}{n!}\right] = e^{1/z}$

$$z\left(\frac{1}{n!}\right) = \sum_{n=0}^{\infty} \frac{1}{n!}\, z^{-n}$$

$$= 1 + \frac{z^{-1}}{1!} + \frac{z^{-2}}{2!} + \frac{z^{-3}}{3!} + \ldots = e^{1/z}$$

Shifting $\left(\dfrac{1}{n!}\right)$ one unit to the left

$$z\left[\frac{1}{(n+1)!}\right] = z\left[z\left(\frac{1}{n!}\right) - 1\right] = z\left(e^{1/z} - 1\right)$$

Shifting $\left(\dfrac{1}{n!}\right)$ two units left

$$z\left[\frac{1}{(n+z)!}\right] = z^2[e^{1/z} - 1 - z^{-1}].$$

Solution:

Note : If a unit impulse occurs at the origin, then the unit impulses sequences defined as

$$f_n = \begin{cases} 1 \text{ for } n = 0 \\ 0 \text{ otherwise} \end{cases}$$

$x\,(f_n) = 1$

The two basic theorems. We often need the values of u_n for n = 0 or as $n \to \infty$ without requiring the complete knowledge of u_n.

We can find this as the behaviour of u_n for small values of n is related to the behaviour of $\overline{u}(z)$ as $z \to \infty$ and vice-versa. The precise relationship is given by the following initial & final value theorems.

Initial value theorems says that

If $z(u_n) = \overline{u}(z)$ then

$$u_0 = \underset{z\to\infty}{\text{Lt}}\ \overline{u}(z)$$

& final value theorem says that

of $z(u_n) = \overline{u}(z)$ then

$$\underset{z\to\infty}{\text{Lt}}\ (u_n) = \underset{z\to\infty}{\text{Lt}}\ (z-1)\ \overline{u}(z)$$

Example 5:

If $\overline{u}(z) = \dfrac{2z^2 + 5z + 14}{(z-1)^4}$ find 42 & 43

$$\overline{u}(z) = \frac{1}{z^2} \cdot \frac{2 + 5z^{-1} + 14z^{-2}}{\left(1 - z^{-1}\right)^4}$$

By initial value theorem

$$u_0 = \underset{z\to\infty}{\text{Lt}}\ \overline{u}(z) = (0)$$

Solution:

Similarly

$$u_1 = \underset{z\to\infty}{\mathrm{Lt}} \left[z\,\overline{u}(z) - u_0\right] = 0$$

$$u_2 = \underset{z\to\infty}{\mathrm{Lt}} \left[z^2\left[\overline{u}(z) - u_0 - u_1 z^{-1}\right]\right]$$

$$= 2 - 0 - 0 = 2$$

$$u_3 = \underset{z\to\infty}{\mathrm{Lt}}\ z^3\left[\overline{u}(z) - u_0 - u_1 z^{-1} - u_2 z^{-2}\right]$$

$$= \underset{z\to\infty}{\mathrm{Lt}}\ z^3\left[\overline{u}(z) - 0 - 0 - 2\,z^{-2}\right]$$

$$= \underset{z\to\infty}{\mathrm{Lt}}\ z^3\left[\frac{2z^2+5z+14}{(z-1)^4} - \frac{2}{z^2}\right]$$

$$= \underset{z\to\infty}{\mathrm{Lt}}\ z^3\left[\frac{13z^3 + 2z^2 + 8z - 2}{z^2(z-1)^4}\right] = 13.$$

Example 6:

A point $z = re^{i\Omega}$ is located at a distance Ω from the origin and angle Ω relative to the real axis. Then the discrete time function transform (DTFT) is obtained from z-transform by setting $r = 1$, or substituting $z = e^{i\Omega}$ in the equation

$$X(z) = \sum_{n=-\infty}^{\infty} x[n]z^{-n} \qquad ...(1)$$

i.e. $X(e^{i\Omega}) = X(z)/z = e^{i\Omega}$

Solution:

The equation $z = e^{iW}$ describes a circle of unit radius centered on the origin in the z plane as shown in fig. This contour is termed as the unit circle in z plane. Frequency W in the DTFT corresponds to a point on the unit circle at an angle of W w.r.t. the real axis.

As discrete time frequency Ω goes from $-\pi$ to π, we take one trip around the unit circle.

Example 7:

Determine the z-transform of $x[n] = -u[-n-1] + \left(\frac{1}{2}\right)^n u[n]$

Comment on ROC & pole-zero location of X(z) in the z-plane.

Solution:

$$X(z) = \sum_{n=-\infty}^{\infty}\left[\left(\frac{1}{2}\right)^n u[n]\, z^{-n} - u[-n-1]\, z^{-n}\right]$$

$$= \sum_{n=-\infty}^{\infty}\left(\frac{1}{2z}\right)^n - \sum_{n=-\infty}^{\infty}\left(\frac{1}{z}\right)^n$$

$$= \sum_{n=\infty}^{\infty}\left(\frac{1}{2z}\right)^n + 1 - \sum_{k=0}^{\infty} z^k$$

Both the sums in this example must converge is order for X(z) to converge. This implies that we must have

$$\left(\frac{1}{2}\right)\left(\frac{1}{z}\right) < 1 \quad |z| < 1$$

or $|z| > \frac{1}{2}$ &

Hence

$$X(z) \Rightarrow \frac{1}{1-\left(\frac{1}{2}\right)z^{-1}} + 1 - \frac{1}{1-z}\ \frac{1}{2} < |z| < |$$

$$= \frac{z\left(2z - \frac{3}{2}\right)}{\left(z-\frac{1}{2}\right)(z-1)}\ \frac{1}{2} < |z| < |$$

ROC is a ring in the z-plane.

Example 8:

$$x[n]\left(\frac{1}{2}\right)^n u[n] + \left(\frac{-1}{3}\right)^n u[n]$$

Answer is $X(z) = \dfrac{z\left(2z-\frac{1}{6}\right)}{\left(z-\frac{1}{2}\right)\left(z+\frac{1}{3}\right)}$

$|z| > \frac{1}{2}$

Solution:

Properties of Region of Convergence

In this, we show how the ROC is related to the characteristics of a signal, x[n]. The ROC properties, we can often identify the ROC from X(z) and limited knowledge of the characteristics of x[n]. The relationship between the ROC and the characteristics of the fine domain signal is used to find the increase z-transform.

First we know that the ROC cannot can train only poles. This is because ROC is defined as the set of z for which the z-transform converges. Hence X(z) must be finite for all z in the ROC. However if d is a pole, then $|x(\alpha)| = \infty$ and the z-transform does not converge at the pole. Thus the pole cannot lie is the ROC.

Second : The ROC for a finite duration signal includes the entire z-plane, except possibly z = 0 and/or $|z| = \infty$. To see this, suppose x[n] is non-zero only on the interval $n_1 \leq n \leq n_2$, We have

$$X(z) = \sum_{n=n_1}^{n_2} x[n]\, z^{-n}$$

This sum will converge provided that each term in the sum in finite.

Now consider infinite-duration signals. The condition for convergence is |X(z)|< ¥ we may thus write

$$|X(z)| = \left| \sum_{n=n-\infty}^{\infty} x[n]\, z^{-n} \right|$$

$$\leq \sum_{n=n-\infty}^{\infty} \left|x[n]\, z^{-n}\right|$$

$$= \sum_{n=n-\infty}^{\infty} |x[n]|\, |z^{-n}|$$

The second line follows from the fact that the magnitude of a sum of numbers is less than or equal to sum of magnitudes.

We obtain the third line by noting that the magnitude of a product equal to the product of magnitudes. Splitting the unifinite sum into negative and positive time portions. We may write

$$X_-(z) = \sum_{n=n-\infty}^{-1} |x(n)|\, |z^{-n}|$$

and $X_+(z) = \sum_{n=0}^{\infty} |x(n)| \, |z^{-n}|$

We note that $|X(z)| \leq X_-(z) + X_+(z)$. If both $X_-(z)$ and $X_+(z)$ are finite, then $|X(z)|$ is guaranteed finite. This clearly shows that $|x[n]|$ be bounded is some way.

Suppose we can bound $|x[n]|$ by finding small positive constants A_-, A_+, r_-, r_+ such that

$|x[n]| \leq A_-(r_-)^n \; n < 0$ ...(1)

and $|x[n]| \leq A_+ (r_+)^n \; n \geq 0$ ...(2)

A signal that satisfies these bounds grows no faster than $(r_+)^n$ for positive n and $(r_-)^n$ for negative n. While we can construct signals that do not satisfy these bounds, such as a^{n^2}, such type of signal do not appear in an engineering problems.

If bound is satisfied then

$$X_-(z) \leq A_- \sum_{n=-\infty}^{\infty} (r_-)^n \, |z^{-n}|$$

$$= A_- \sum_{n=-\infty}^{-1} \left(\frac{r_-}{|z|}\right)^n$$

$$= A_- \sum_{n=1}^{\infty} \left(\frac{|z|}{r_-}\right)^k$$

This last sum converges if and only if

$|z| < r_-$. Now consider positive time function

$$X_+(z) \leq A_+ \sum_{n=0}^{\infty} (r_+)^n \, |z|^{-n}$$

$$A_+ \sum_{n=0}^{\infty} \left(\frac{r_+}{|z|}\right)^n$$

This sum converges if and only if $|z| > r_+$. Hence if $r_+ < |z| < r_-$ then both $X_+(z)$ and $X_-(z)$ converges and $|X(z)|$ also converges. Note that if $r_+ > r_-$ then there are no values of z for which convergence is guaranteed.

This analyses supports the following conclusions for infinite duration signals. Define self-sided signal as one for which $x[n] = 0$ for $n > n_2$ and right sided signal as one for which $x[n] = 0$ for $n < n$, and two sided signal a sone that has finite value (duration) is both positive and negative directions. For signals x[n] that satisfy the experiential bounds in equation 1 & 2 we have

1. The ROC of a right sided signal is of the form $|z| > r_+$.
2. The ROC of a left sided signal is of the form $|z| < r_-$.
3. The ROC of a two sided signal is of the from
   $r_+ < |z| < r_-$

One sided experiential signals are frequently encountered in engineering problem. In this case the magnitude of one or more poles determines the ROC boundaries r_- and r_+.

Suppose we have a right handed signal $x[n] = \alpha^n u[n]$ The z-transform of x[n] has a pole at $z = \alpha$ and the ROC is $|z| > |\alpha|$ Hence the ROC is the region of the z plane with a radius greater than the radius of pole. Like wise if $x[n] = \alpha^n u[-n-1]$ (a left sides f^n), the ROC is $|z| < |\alpha|$; the region of the z-plane with radius less than the radius of the pole.

Example 9:

Identify the ROC associated with the z-transform for each of the following

$$x[n] = \left(-\frac{1}{2}\right)^n u[-n] + 2\left(\frac{1}{4}\right)^n u[n]$$

$$y[n] = \left(-\frac{1}{2}\right)^n u[n] + 2\left(\frac{1}{4}\right)^n u[n]$$

$$w[n] = \left(-\frac{1}{2}\right)^n u[-n] + 2\left(\frac{1}{4}\right)^n u[-n]$$

Solution

$$X(z) = \sum_{-\infty}^{0}\left(-\frac{1}{2z}\right)^n + 2\sum_{n=0}^{\infty}\left(\frac{1}{4z}\right)^n$$

$$\Rightarrow \sum_{n=0}^{\infty}(-2z)^k + 2\sum_{n=0}^{\infty}\left(\frac{1}{4z}\right)^n$$

The first series converges for $|z| < \frac{1}{2}$ while second converges for $|z| >$ 1/4 both series must converge for X(z) to converge, so the ROC is

$\frac{1}{4} < |z| < \frac{1}{2}$ & we set

$$X(z) \Rightarrow \frac{1}{1+2z} + \frac{2z}{z - \frac{1}{4}}$$

which has a pole at $z = \frac{-1}{2} + z = \frac{1}{4}$. The ROC is in the ring shaped region located between the poles.

$$Y(z) \Rightarrow \sum_{n=0}^{\infty} \left(\frac{-1}{2z}\right)^n + 2\sum_{n=0}^{\infty} \left(\frac{1}{4z}\right)^n$$

First series converges for

$|z| < \frac{1}{2}$ while second $|z| > \frac{1}{4}$ Hence the combined ROC is $|z| < \frac{1}{2}$

$$Y(z) = \frac{z}{z + \frac{1}{2}} + \frac{2z}{z - \frac{1}{4}}$$

for which the poles at

$$z = \frac{-1}{2} \ \& \ z = \frac{1}{4}.$$

$$W(z) = \sum_{n=-\infty}^{\infty} \left(\frac{-1}{2z}\right)^n + 2\sum_{n=-\infty}^{0} \left(\frac{1}{4z}\right)^n$$

$$= \sum_{n=\infty}^{\infty} (-2z)^n + 2\sum_{n=0}^{\infty} (uZ)^n$$

Here the first series converges for

$|2z| < 1$ i.e. $|z| < \frac{1}{2}$

while the second $|z| < \frac{1}{4}$ & the combined is $|z| < \frac{1}{4}$

$$W(z) \Rightarrow \frac{1}{1 + 2z} + \frac{2}{1 - 4z}$$

where poles at $z = \frac{-1}{2}$ & $z = \frac{1}{4}$.

The above example shows that ROC of a two sides signal is a ring, the ROC of right sided is extension of a circle & the ROC of left sided signal is the interior of a circle & poles defines the boundaries.

Example 10:

Determine z-transform and ROC of two sided signal $x[n] = z^{|n|}$ assuming $|\alpha| < 1$. Repeat for $|\alpha| > 1$.

Solution:

For $|\alpha| < 1$

$$X(z) = \frac{z}{z-\alpha} - \frac{z}{z-\frac{1}{\alpha}} \quad |\alpha| < |z| < \frac{1}{|\alpha|}$$

for $|a| > 1$, the ROC is empty set.

Properties of z-transform

1. **Linearity :** The linearity properly says that the z-transform of a sum of signals is just the sum of the individual z-transform i.e.

$a\,x[n] + by\,[n] \xleftarrow{z} a\,X(z) + b\,Y(z)$ with ROC is the intersection of individual ROC's because the z-transform of the sum is valid only where both X(z) & Y(z) converges. The ROC can be larger than the intersection if one more terms in x[n] or y[n] cancels each other in the sum. In the z-plane this corresponds to a zero cancelling a pole that defines one of the ROC boundaries.

Example 11:

Suppose

$x[n] = \left(\frac{1}{2}\right)^n u[n] - \left(\frac{3}{2}\right)^n u(-n-1)$ Its z-transform is

$$\frac{-z}{\left(z-\frac{1}{2}\right)\left(z-\frac{3}{2}\right)} \text{ with ROC}$$

$\frac{1}{2} < |z| < \frac{3}{2}$ and $y[n] = \left(\frac{1}{4}\right)^n u(n) - \left(\frac{1}{2}\right)^n u[n]$

$$\xleftrightarrow{n} Y(z) = \frac{-\frac{1}{4}z}{\left(z-\frac{1}{4}\right)\left(z-\frac{1}{2}\right)} \text{ with ROC } |z| > \frac{1}{2}.$$

Evaluate z-transform of a x[n] + b y[n]

Solution:

The pole zero plot of X(z) & Y(z) is as shown

Linearity property indicates that

a x[n] + b y[n]

$$\xleftrightarrow{z} a\frac{-z}{\left(z-\frac{1}{2}\right)\left(z-\frac{3}{2}\right)}+b\frac{-\frac{1}{4}z}{\left(z-\frac{1}{4}\right)\left(z-\frac{1}{2}\right)}$$

In general the ROC is the intersection of individual ROC's or $\frac{1}{2}<z<\frac{3}{2}$ which corresponds to ROC of X(z) note however, what happens when a = b in this case we have

a x[n] + a y[n]

$$= a\left(-\left(\frac{3}{2}\right)^n u[-n-1]+\left(\frac{1}{4}\right)^n u[n]\right)$$

we see that term $\left(\frac{1}{2}\right)^n u[n]$ has been cancelled in the time domain signal. The ROC is now easily verified to be $\frac{1}{4}<z<\frac{3}{2}$.

The ROC is larger than the intersection of the individual ROC because the term $\left(\frac{1}{2}\right)^n u[n]$ is no larger present. Combining z-transforms and using linearity properly a X(z) + a Y(z) =

$$a\left[\frac{-z}{\left(z-\frac{1}{2}\right)\left(z-\frac{3}{2}\right)}+\frac{-\frac{1}{4}z}{\left(z-\frac{1}{4}\right)\left(z-\frac{1}{2}\right)}\right]$$

$$\Rightarrow a\frac{-\frac{1}{4}z\left(z-\frac{3}{2}\right)-z\left(z-\frac{1}{4}\right)}{\left(z-\frac{1}{4}\right)\left(z-\frac{1}{2}\right)\left(z-\frac{3}{2}\right)}$$

$$\Rightarrow \frac{-\frac{5}{4}z\left(z-\frac{1}{2}\right)}{\left(z-\frac{1}{4}\right)\left(z-\frac{1}{2}\right)\left(z-\frac{3}{2}\right)}$$

The zero at $z=\frac{1}{2}$ cancels the pole at $z=\frac{1}{2}$ so

$$a\,X(z) + a\,Y(z) = a\frac{-\frac{5}{4}z}{\left(z-\frac{1}{4}\right)\left(z-\frac{3}{2}\right)}$$

Hence cancellation of the $\left(\frac{1}{2}\right)^n u[n]$ term in the time domain corresponds to the cancellation of the pole at $z = \frac{1}{2}$ by a zero in the z-domain. This poled defined the ROC boundary, so the ROC enlarges when it is removed.

Property

Time Reversal $x[-n] \xleftrightarrow{z} X\left(\frac{1}{z}\right)$ with ROC $\frac{1}{R_x}$ Time reversal or reflection, corresponds to replacing z by z^{-1}. Hence if R_x is of the form a < |z| < b, the ROC of the reflected signal is

$a < 1/|z| < b$ or $\frac{1}{b} < z < \frac{1}{a}$.

Time Shift

$$x[n - n_0] \xleftrightarrow{z} z^{-n_0} X(z)$$

with ROC R_x except possibly z = 0 or |z| = ¥ multiplication by z^{-n_0} introduces a pole of order n_0 at z = 0 if $n_0 > 0$. If is this case the ROC can not include z = 0, even if R_x does include z = 0, unless X(z) has a zero of at least order n_0 at z = 0 that cancels all of the new poles. If $n_0 < 0$, then multiplication by z^{-n_0} introduces n_0 poles at infinity.

If these poles are not cancelled by zeros at infinity in X(z), then ROC of z^{-n_0} X(z) can not include |z| = ¥

Multiplication by Exponential Sequence

$$\alpha^n x[n] \xleftrightarrow{z} X\left(\frac{z}{\alpha}\right)$$

Convolution

$$x[n] + y[n] \xleftrightarrow{z} X(z)\,Y(z)$$

Convolution of time signals corresponds to multiplication of z-transform. As in the linearity property, the ROC may be larger than the intersection if a pole-zero cancellation occurs in the product X(z) Y(z).

Differentiation in the z-domain

$$n\, x[n] \overset{z}{\longleftrightarrow} -z \frac{d}{dz} X(z)$$

multiplication by n in time domain corresponds to differentiation with respect to z and multiplication of the result by –z in the z domain. This operation does not charge the ROC.

Example 12:

Find z-transform of signal

$$x[n] = \left(n\left(\frac{-1}{2}\right)^n u[n]\right) \pm \left(\left(\frac{1}{4}\right)^{-n} u[-n]\right).$$

Solution:

First we find out the z-transform of

$$w[n] = n\left(\frac{-1}{2}\right)^n u[n]$$

$$\left(\frac{-1}{2}\right)^n u[n] \overset{z}{\longleftarrow} \frac{z}{z+\frac{1}{2}} \text{ with ROC} > \frac{1}{2}$$

Thus the z domain differentiation properly implies

$$w[n] = n\left(\frac{-1}{2}\right)^n u[n]$$

$$\overset{z}{\longleftarrow} W(z) = -z\frac{d}{dz}\left(\frac{z}{z+\frac{1}{2}}\right) \text{with ROC}|z|>1/2$$

$$\Rightarrow -z\left[\frac{z+\frac{1}{2}-z}{\left(z+\frac{1}{2}\right)^2}\right]$$

$$\Rightarrow \frac{\frac{-1}{2}z}{\left(z+\frac{1}{2}\right)^2} \text{ with ROC } |z| > \frac{1}{2}$$

Next we find the z-transform of $y[n] = \left(\frac{1}{4}\right)^{-n} u[-n]$. We do this by using time reversal property which says $x[-n] \xleftarrow{z} X\left(\frac{1}{2}\right)$ with ROC & $\frac{1}{R_x}$

$$\left(\frac{1}{u}\right)^n u[+n] \xleftarrow{z} \frac{z}{z-\frac{1}{4}} \quad \text{with ROC } |z| > \frac{1}{4} \quad \text{hence}$$

$$y[n] \xleftarrow{z} Y(z) \Rightarrow \frac{z^{-1}}{z^{-1}-\frac{1}{4}} \text{ with ROC } \frac{1}{|z|} > \frac{1}{4}$$

$$\Rightarrow \frac{-4z}{z-z} \text{ with ROC } |z| < 4$$

Lastly we apply the convolution property $x[n] = w[n] + y[n] \xleftarrow{z}$ $X(z) = W(z)\,Y(z)$

$$\Rightarrow \frac{\frac{-1}{2}z}{\left(z+\frac{1}{2}\right)^2} + \frac{-4z}{(z-4)}\ \frac{1}{2} < |z| < 4$$

$$\Rightarrow \frac{2z^2}{(z-4)\left(z+\frac{1}{2}\right)^2}\ \frac{1}{2} < |z| < 4$$

Inversion of z-transform

There are two methods for determining inverse z-transform.

1. The method of partial fractions uses knowledge of several basic z-transform pairs and the z-transform properties to invert a large class of z-transform. This approach also relies on an important property of the ROC. i.e. a right sides time signal has a ROC that lies outside the pole radius & left sided time signal has a ROC that lies inside the pole radius.
2. The second inversion method express X(z) as a power series in Z^{-1} of the form

$$X(z) = \sum_{n=-\infty}^{\infty} x[n]\, z^{-n}$$

So that the values of signal may be determined by inspection

$$X(z) = \frac{2z^2 - 2z + 10}{3z^3 - 6z + 9}$$

then we factor z^2 from numerator & $3z^3$ from denominator as

$$X(z) = \frac{z^2}{3z^3}\left(\frac{2 - 2z^{-1} + 10z^{-2}}{1 - 2z^{-2} + 3z^{-3}}\right)$$

$$= \frac{1}{3}z^{-1}\left(\frac{2 - 2z^{-1} + 10z^{-2}}{1 - 2z^{-2} + 3z^{-3}}\right)$$

The partial fraction expansion is of equation (1) is obtained by factoring the denominator polynomial into a product of first orter terms as

$$X(z) = \frac{b_0 + b_1 z^{-1} + \ldots + b_M z^{-M}}{\pi_{k=1}^{N}\left(1 - d_x z^{-1}\right)}$$

Where d_x are the poles of X(z). If none of the poles are repeated, then we may rewrite X(z) as a sum of first-order terms using the partial fractions expansion 1

$$X(z) = \sum_{n=1}^{N} \frac{A_k}{1 - d_k z^{-1}}$$

The inverse z-transform associated with each term is then determined using the transform pair.

Partial Fraction Expansions

In the study of LTI system, we often encounter z-transform that are a rational function of z^{-1}

$$X(z) = \frac{B(z)}{A(z)}$$

$$== \frac{b_0 + b_1 z^{-1} + \ldots + b_M z^{-M}}{1 + z_1 z^{-1} + \ldots + a_N z^{-N}}$$ and assu-me that $M < N$. If $M \geq N$, the we may use long division to express X(z), in the form

$$X(z) = \sum_{k=0}^{M-N} f_k z^{-k} + \frac{B^2(z)}{A(z)}$$

The numerator polynomial $\tilde{B}$ (z) now has order one less than that of denominator polynomial, and the partial fraction expansion method is applied to determine the inverse transform of $\tilde{B}$ (z)/A(z). The inverse z-transform of the term in the sum are obtained from the pair f(n) $\xleftarrow{z}$ 1 and time shifted property.

In some problems x(z) may be expressed as a ratio of polynomials is z rather than z^{-1}. In this case we may use the partial fraction expansion method described here if we first convert x(z) to a ratio of polynomials is z^{-1}.

$$A_k (d_k)^n u[n] \xleftarrow{z} \frac{A_k}{1 - d_k z^{-1}} \text{ with ROC } |z| > d_k$$

or

$$- A_k (d_k)^n u(-n - 1) \xleftarrow{z} \frac{A_k}{1 - d_k z^{-1}} \text{ with ROC } |z| < d_k$$

The relationship between the ROC associated $|z| < d_x$ with X(z) and each pole determines whether the signal is right sided or left sided inverse transform is chosen for each term.

Condition : If a pole di, is repeated r times, then there are r terms in the partial fraction expansion associated with this pole. They are

$$\frac{Ai_1}{1 - di\ z^{-1}}, \frac{Ai_2}{\left(1 - di\ z^{-1}\right)^2}, \ldots \frac{Air}{\left(1 - di\ z^{-1}\right)^r}$$

The z-transform of each term there is determined using either

$$A \frac{(n + 1) \ldots (n + m - 1)}{(m - 1)!} (di)^n u[n]$$

$$\xleftarrow{z} \frac{A}{(1 - di)^m} \text{ with ROC } |z| > di$$

or

$$- A \frac{(n + 1) \ldots (n + m - 1)}{(m - 1)!} (di)^n u(-n - 1)$$

$$\xleftarrow{z} \frac{A}{(1 - di)^m} \text{ with ROC } |z| < di$$

Again the ROC of X(z) determines whether the right or lift sided inverse transform is chosen.

The linearity property indicates that the ROC of X(z) is the intersection cf the ROCs associated with the individual terms in the partial fraction expansion.

In order to choose the correct inverse transform, we must infer the ROC of each term from the ROC of each term from the ROC of X(z). This accomplished by comparing each pole location to the ROC of X(z). If the ROC of X(z) has radius greater than the pole associated with the given terms, we choose right sided inverse transform if the ROC of X(z) has radius less than that of pole we choose left sided inverse transform for this term.

Example 13:

Find the inverse transform of

$$X(z) = \frac{1 - z^{-1} + z^{-2}}{\left(1 - \frac{1}{2}z^{-1}\right)\left(1 - 2z^{-1}\right)\left(1 - z^{-1}\right)}$$ with ROC |<|z| < 2.

Solution:

Use partial fractions expansion to write

$$X(z) = \frac{A_1}{\left(1 - \frac{1}{\alpha}z^{-1}\right)} + \frac{A_2}{\left(1 - 2\,z^{-1}\right)} + \frac{A_3}{\left(1 - z^{-1}\right)}$$

$$= \frac{A_1\left(1-2z^{-1}\right)\left(1-z^{-1}\right) + A_2\left(1-\frac{1}{\alpha}z^{-1}\right)\left(1-z^{-1}\right) + A_3\left(1-\frac{1}{\alpha}z^{-1}\right)\left(1-2z^{-1}\right)}{\left(1-\frac{1}{\alpha}z^{-1}\right)\left(1-2z^{-1}\right)\left(1-z^{-1}\right)}$$

$$1 - z^{-1} + z^{-2} \text{ Þ } A_1(1 - z^{-1} - 2z^{-1} + 2z^{-2}) + A_2\left(1 - z^{-1} - \frac{1}{2}z^{-1} + \frac{1}{2}z^{-2}\right) + A_3\left(1 - 2z^{-1} + \frac{1}{2}z^{-1} + z^{-2}\right)$$

$$\Rightarrow A_1(1 - 3z^{-1}) + 2z^{-2}) + A_2\left(1 - \frac{2}{\alpha}z^{-1} + \frac{1}{2}z^{-2}\right)$$

$$+ A_3\left(1 - \frac{5}{\alpha}z^{-1} + z^{-2}\right)$$

Given

$$\Rightarrow A_1 + A_2 + A_3 = 1$$

$$A_1 + A_2 + A_3 = 1$$

$$+ 3A_1 + \frac{3}{2} A_2 + \frac{5}{2} A_3 = +1$$

$$2A_1 + \frac{1}{2} A_2 + A_3 = 1$$

$$A_1 = 1 \; A_2 = 2 \; A_3 = -2$$

$$X(z) = \frac{1}{1 - \frac{1}{\alpha} z^{-1}} + \frac{2}{1 - 2z^{-1}} - \frac{2}{1 - z^{-1}}$$

Now we find z-transform of each term using the relation between the pole location & the ROC of X(z).

The ROC has the radius greater than z = 1/2, so this term is the right sided inverse transform

$$\left(\frac{1}{2}\right)^n u[n] \xleftarrow{z} \frac{1}{1 - \frac{1}{2} z^{-1}}$$

The ROC has the radius less than the pole at z = 2 so this term is left sided signal

$$-2\,(2)^n\, u(-n-1) \xleftarrow{z} \frac{2}{1 - 2z^{-1}}$$

Last ROC has radius greater than the pole at z = 1 ∴ has right sided inverse of z-transform

$$-2u[n] \xleftarrow{z} \frac{2}{1 - 2z^{-1}}$$

$$\text{So } x[n] = \left(\frac{1}{2}\right)^n u[n] - 2\,(2^n)\, u[-n-1] - 2u[n].$$

Example 14:

Find inverse z-transform of

$$X(z) = \frac{z^3 - 10z^2 - 4z + 4}{2z^2 - 2z - 4} \quad \text{with ROC } |z| <|.$$

Solution:

For poles $2[z^2 - z - 2] \Rightarrow 2[z^2 + z - 2z - 2]$

$\Rightarrow 2[(z)(z + 1) - 2(z + 1)]$

$\Rightarrow 2[(z + 1)(z - 2)]$

i.e. having poles at $z = -1$ & $z = 2$

Now convert X(z) into a ratio of polynomials is z^{-1} acc. to equation

$$X(z) = \frac{B(z)}{A(z)} = \frac{b_0 + b_1 z^{-1} + \ldots + b_M z^{-M}}{1 + a_1 z^{-1} + \ldots + a_N z^{-N}}$$

taking z^3 from numerator & $2z^2$ from denominator

$$X(z) = \frac{1}{2} z \frac{\left(1 - 10z^{-1} - 4z^{-2} + 4z^{-3}\right)}{\left(1 - z^{-1} - 2z^{-2}\right)}$$

The factor $\frac{1}{2}z$ is easily incorporated later using the time shift properly. Using division to reduce the order of numerator polynomial

$$- 2z^{-2} - z^{-1} + 1 \;\overline{\big)\, 4z^{-3} - 4z^{2} - 10z^{-1} + 1}\quad \text{quotient: } -2z^{-1} + 3$$

$$4z^{-3} + 2z^{-2} - 2z^{-1}$$

$$- \quad - \quad +$$

$$-6z^{2} - 8z^{-1} + 1$$

$$-6z^{2} - 3z^{-1} + 3$$

$$+ \quad + \quad -$$

$$-5z^{-1} - 2$$

Thus

$$\frac{1 - 10z^{-1} - 4z^{-2} + 4z^{-3}}{1 - z^{-1} - 2z^{-2}} = -2z^{-1} + 3 + \frac{-5z^{-1} - 2}{1 - z^{-1} - 2z^{-2}}$$

$$\Rightarrow -2z^{-1} + 3 + \frac{-5z^{-1} - 2}{\left(1 + z^{-1}\right)\left(1 - 2z^{-1}\right)}$$

Using partial fractions

$$\frac{-5z^{-1} - 2}{\left(1 + z^{-1}\right)\left(1 - 2z^{-1}\right)} \Rightarrow \frac{A}{\left(1 + z^{-1}\right)} + \frac{B}{\left(1 - 2z^{-1}\right)}$$

$$\Rightarrow \frac{1}{1 + z^{-1}} - \frac{3}{1 - 2z^{-1}}$$

& thus define $X(z) = \frac{1}{2} z\, W(z)$

$$W(z) \Rightarrow -2z^{-1} + 3 + \frac{1}{1+z^{-1}} - \frac{3}{1-2z^{-1}} \text{ with ROC } |z| <|$$

The ROC has a smaller radius then either pole & so the inverse z-transform of W(z) is

$W[n] = -2f[n-1] + 3f[n] - (-1)^n u(-n-1) + 3\ (2)^n u[-n-1]$

The time shift property indicates

$$x[n] = \frac{1}{2} w[n+1]$$

Thus

$$x[n] = -\infty[n] + \frac{3}{2} f[n+1] - \frac{1}{2}(-1)^{n+1} u[-n-2] + \frac{3}{2}(2)^{n+1} u[-n-2]$$

Power Series Expansion

Here we seek to express X(z) as a power series is z^{-1} or z of the form

$$X(z) = \sum_{n=-\infty}^{\infty} x[n]\, z^{n-n}$$

The value of the signal x[n] is then given by the coefficients associated with z^{-n}. This inversion method is limited to signals that are on sided; that is, with ROCs of the form $|z| < a$ or $|z| > a$. If the ROC is $|z| > a$, then we express X(z) as a power series is z^{-1} so that we obtain a right sided signal. If ROC is $|z| < a$, then we express X(z) as a power series is z and obtain a left sided inverse transform.

An advantage of the power series approach is the ability to find z. inverse z-transforms for signals that are not a ratio of polynomials is z.

Example 15:

Find inverse z-transform of

$$X(z) = \frac{2+z^{-1}}{1-\frac{1}{2}z^{-1}} \text{ with ROC } |z| > \frac{1}{2}$$

using a power series expansion.

Solution:

Use long division to write X(z) as a power series in z^{-1}, since the ROC indicates that x[n] is right sided we have

$$1-\frac{1}{2}z^{-1}\ \overline{\sqrt{2+z^{-1}}}\quad 2+2z^{-1}+z^{-2}+\frac{1}{2}z^{-2}+\ldots$$

$$2-z^{-1}$$

$$\underline{-\ +}$$

$$2z^{-1}$$

$$\underline{2z^{-1}-z^{-2}}$$

$$z^{-2}$$

$$\underline{z^{-2}-\frac{1}{2}z^{-3}}$$

$$\frac{1}{\alpha}z^{-3}$$

i.e.

$$X(z) = 2 + 2z^{-1} + z^{-2} + \frac{1}{2}z^{-3} + \ldots$$

thus we conclude

$x[n] = = 0 \qquad n < 0$

$x[0] = 2$

$x[1] = 2$

$x[2] = 1$

$x[3] = \frac{1}{2}$ ...

If the ROC is changed to $|z| < \frac{1}{2}$, less we expand X(z) as power series is z.

$$\frac{-1}{\alpha}z^{-1}+1\ \overline{z^{-1}+2}\quad -2-8z-16z^2-32z^3+\ldots$$

$$\underline{z^{-1}-2}$$

$$4$$

$$\underline{4-8z}$$

$$8z$$

$$\underline{8z-16z^2}$$

$$16z^2$$

$$X(z) = -2 - 8z - 16z^2 - 32z^3 + \ldots$$

thus x[n] = 0 n > 0

x[0] = – 2

x[–1] = – 8

x[–2] = – 16

x[–3] = – 32

Example 16:

Find this inverse z-transform of

$X(z) = e^{z^2}$ with ROC all z except $|z| = \infty$.

Solution

Using power series expansion

$$e^a = \sum_{n=0}^{\infty} \frac{a^k}{k!}$$

$$\text{we write } X(z) = \sum_{n=0}^{\infty} \frac{\left(z^2\right)^n}{n!}$$

$$= \sum_{n=0}^{\infty} \frac{z^{2n}}{k!}$$

$$\text{thus } x[n] = \begin{cases} 0 & n > 0 \text{ and } n \text{ add} \\ \dfrac{1}{\left(\dfrac{n}{2}\right)!} & \text{otherwise} \end{cases}$$

Stability & Casualty [Causality]

The impulse response of a casual system is zero for n < 0. Therefore, the impulse response of a casual system is determined from the T.F. by using right sided inverse transform. A pole that is inside the unit circle is the z-plane ($|d_x|<|$) contributes an exponentially decaying term to the impulse response, while a pole that is outside the unit circle ($|d_x| > 1$) contributes am exponentially increasing term.

Alternatively, if we know system is stable, then impulse response is absolutely summable and the DTFT of the impulse response exists. It follows that, the ROC must include unit circle in the z plane. Hence the relationship between a pole location and the unit circle determines the component of the impulse response associated with the pole. The pole inside the unit circle contributes a right sided decaying exponential term to the impulse response, while a pole outside the unit circle contributes a left sided decaying exponential term to the impulse response.

System that are both stable & casual must have all their poles inside the unit circle. A pole that is inside the unit circle is the z plane contributes a right sided or casual decaying exponential term to the impulse response. We cannot have pole outside the unit circle, since the inverse transform of a pole located outside the circle will contribute other a right-sided increasing exponential term, which is not stable, or a left sided decaying exponential, with is not casual.

Example of stable & casual system is

Transform Analysis of LTI System

The output y[n] of a LTI system is expressed in terms of the impulse response h[n] & the input x[n] using convolution

$$y[n] = h[n] + x[n]$$

We take z-transform on both side

$$Y(z) = H(z)\, X(z)$$

the z-transform has converted convolution of time sequences into multiplication of transforms. Thus the transfer function offers another description of the o/p-o/p char. of a LTI discrete time system.

The T.F. may also the viewed as the ratio of z-transform of the output to that of input i.e.

$$H(z) = \frac{Y(z)}{X(z)}$$

This defines for all values of z for which X(z) is more zone.

The Relationship between T.F. & Difference Equation

The transfer function may be obtained directly from the difference equation description of a system. The difference equation relates o/p y[n] & o/p x[n] as

$$\sum_{n=0}^{N} a_k \, y[n-k] = \sum_{n=0}^{M} b_k \, x[n-k]$$

We know that the transfer f^n. H(z) is an eigen value of the system associated with the eigen function z^n. That is, if $x[n] = z^n$, then the o/p of LTI system is $y[n] = z^n H(z)$.

Substituting $x[n-k] = z^{n-k}$ and $y[n-k] = z^{n-k} H(z)$ into the above relation

$$z^n \sum_{n=0}^{N} a_k z^{-k} H(z) = z^n \sum_{n=0}^{M} b_k z^{-k}$$

$$H(z) = \sum_{n=0}^{M} b_k z^{-k}$$

$$H(z) = \frac{\sum_{n=0}^{M} b_k z^{-k}}{\sum_{n=0}^{N} a_k z^{-k}} \Rightarrow$$

The transfer f^n of a system described by a difference equation is a ratio of polynomials is z^{-1} and is thus termed as rational transfer f^n the coefficient z^{-k} is the numerator is associated with $x[n-k]$ is the difference equation. The coefficient of z^{-k} is the denominator polynomial is the coefficient associated with $y[n-k]$ is the difference equation.

We can now find out the transfer function given the difference equation. But also to find as difference equation description for the system given a rational T.F.

The T.F. is pole zero form by factory the numerator & denominator in the above equation

$$\frac{\tilde{b}\pi_{n=1}^{M}\left(1-\left(x\, z^{-1}\right)\right)}{\pi_{n=1}^{N}\left(1-d_x\, z^{-1}\right)}$$

where $\tilde{b} = b_p/a_1$.

The impulse response is the inverse z-transform of the transfer equation. However difference equation does not provide any ROC information; so other system characteristic, such as stability or causality, must be known in order to uniquely determine impulse response.

Example 17:

Determine the z-transform of the signal

$y[n] = -\alpha^n u[-n-1]$

Comment on ROC and the pole zero location in z-plane.

Solution:

Substitute $y[n]$ into the equation

$$X(z) = \sum_{n=-\infty}^{\infty} x[n]\, z^{-n}$$

$$X(z) = \sum_{n=-\infty}^{\infty} -\alpha^{n}\, u[-n-1]\, z^{-n}$$

$$= - \sum_{n=-\infty}^{-1} \left(\frac{\alpha}{z}\right)^{n}$$

$$\Rightarrow - \sum_{k=1}^{\infty} \left(\frac{z}{\alpha}\right)^{k}$$

$$\Rightarrow 1 - \sum_{k=1}^{\infty} \left(\frac{z}{\alpha}\right)^{k}$$

The sum converges for $\left|\frac{z}{\alpha}\right| < \Big|$ or $|z| < |\alpha|$

Hence $Y(z) = 1 - \dfrac{1}{1 - z\,\alpha^{-1}} \quad |z| < |\alpha|$

$$= \frac{z}{z-\alpha} \quad |z| < |\alpha|$$

Note that Y(a) is identical to X(z) even though the time signals are quite different. Only the ROC differentiates them. We must know the ROC to determine the correct inverse z-transform.

Example 18:

Find the difference equation description for a system with T.F.

$$H(z) = \frac{5z + 2}{z^2 + 3z + 2}.$$

Solution:

First write H(z) as a polynomial is z^{-1}. Dividing both the numerator & denominator by z^2

$$H(z) = \frac{5z^{-1} + 2z^{-2}}{1 + 3z^{-1} + 2z^{-2}}$$

Comparing this with $H(z) = \dfrac{\sum_{n=0}^{M} b_k\, z^{-k}}{\sum_{n=0}^{N} a_k\, z^{-k}}$

We conclude that M = 2, N = 2, $b_0 = 0$, $b_1 = 5$, $b_2 = 2$, $a_0 = 1$, $a_1 = 3$, $a_2 = 2$. Hence this system is described by difference equation

$$\sum_{n=0}^{N} a_k \; y[n-k] = \sum_{n=0}^{M} b_k \; x[n-k]$$

a_0 y[n] + 3y [n – 1] + 2y [n – 2]

⇒ 5x[n – 1] + 2x[n – 2].

Problem

A system is described by a difference equation

$$y[n] - y[n-1] + \frac{1}{4} y[n-2 = x[n] + \frac{1}{4} x[n-1] - \frac{1}{8} x[n-2]$$

Find T.F. of inverse. Does the system is stable & casual.

Solution:

$$H(z) \Rightarrow \frac{\sum_{n=0}^{M} b_k \; z^{-1}}{\sum_{n=0}^{N} a_k \; z^{-k}}$$

$$= \frac{1 + \frac{1}{4} z^{-1} - \frac{1}{8} z^{-2}}{1 - z^{-1} + \frac{1}{4} z^{-2}}$$

$$H(z) = \frac{\left(1 - \frac{1}{4} z^{-1}\right)\left(1 + \frac{1}{2} z^{-1}\right)}{\left(1 - \frac{1}{2} z^{-1}\right)^2}$$

$$H^{-1}(z) \Rightarrow \frac{\left(1 - \frac{1}{2} z^{-1}\right)^2}{\left(1 - \frac{1}{4} z^{-1}\right)\left(1 + \frac{1}{\alpha} z^{-1}\right)}$$

Since the poles of the inverse system are at $z = \frac{1}{4}$ & $z = \frac{-1}{2}$. Both of these poles are inside the unit circle & ∴ the system can be both stable & casual.

3

Nyquist Stability Criterion

NTRODUCTION

Nyquist stability criterion relates the location of roots of the characteristic eq. to the open-loop frequency response. The computation of closed-loop poles is not necessary and the stability study can be carried out graphically from the open-loop frequency response. The Nyquist stability criterion is based on the principle of argument which is given below.

Enclosement and Encirclement

A point or region is said to be 'enclosed' by a closed path if it is found to lie to the right of the path when path is traversed in the clock-wise direction. A point is said to be 'encircled' by a closed-path if it is found inside the path.

NYQUIST CONTOUR

A contour which completely encloses right half of the s-plane is called as **Nyquist contour.** It is directed clockwise and comprises of an infinite line segment C_1 along jω axis and an arc C_2 of infinite radius.

Along C_1; $s = j\omega$ with 's' varying from $-j\infty$ to $+j\infty$; and

along C_2; $s = \underset{R\to\infty}{R_e^{j\theta}}$ with 'θ' varying from $+\frac{\pi}{2}$ to $-\frac{\pi}{2}$.

The Nyquist contour so defined encloses all the right half s-plane poles and zeros of $q(s) = 1 + G(s) . H(s)$

Nyquist Stability Criterion

If the contour T_{GH} of the open-loop T.F. G(s) . H(s) corresponding to the Nyquist contour in the S-plane encircles the point $[-1 + j0]$ in the counter clockwise direction as many times as the number of right half s-plane poles of G(s) . H(s), then the closed-loop system is stable.

Principle of Argument

Let $q(s) = \dfrac{(s-\alpha_1)(s-\alpha_2)....(s-\alpha_m)}{(s-\beta_1)(s-\beta_2)....(s-\beta_n)}$ be a polynomial;

Where s is a complex variable; represented by $s = \sigma + j\omega$ in complex s-plane. Then the function q(s) must also be complex and may be defined as $q(s) = u + jv$ and represented on the complex plane q(s). Hence for every point s in s-plane at which q(s) is analytical, there exist a point q(s) in q(s) plane.

Therefore for a contour in the s-plane which does not go through any singular point, there corresponds a contour in the q(s) plane. If there are 'P' poles and 'Z' zeros of q(s) enclosed by the s-plane contour then the corresponding q(s) plane contour must encircle the origin 'Z' times in clockwise direction and 'P' times in counter clockwise direction, resulting in a net encirclement of the origin '[P – Z]' times in the counter clockwise direction. This relation between the enclosure of poles and zeros of q(s) by the s-plane contour and the encirclement of the origin by the q(s) plane contour is known as principle of argument.

Mapping of the Nyquist contour into the contour T_{GH}

(1) The mapping of the imaginary axis is carried out by substitution of $s = j\,\omega$ in G(s).H(s). This converts the mapping function into a frequency function of $G(j\omega)\,.\,H(j\omega)$.

(2) In physical system $(m \le n)$; $\lim_{\substack{s = Re^{j\theta} \\ R\to\infty}} G(s)\,.\,H(s) = \text{Real constant}$

$= 0$ if $m < n$

Thus the infinite arc of the Nyquist contour maps into a point on the real axis.

Therefore the complete contour T_{GH} is the polar plot of $G(j\omega)\,.\,H(j\omega)$ with ω varying from $-\infty$ to ∞. This is usually called *Nyquist Plot.* Nyquist plot is symmetrical about the real axis. Hence if polar plot of $G(j\omega)\,.\,H(j\omega)$ for ω from 0 to ∞ is plotted and to this plot its mirror image about the real axis is joined we will get the Nyquist Plot.

(3) When open-loop poles lie on $j\,\omega$ axis : Mapping of Nyquist contour is accomplished by indenting the Nyquist contour around the $j\omega$ – axis poles along a semicircle of radius ε, where $\varepsilon \to 0$ as shown below.

It has total four parts as under :

Part I : Complete $j\,\omega$ - axis :

It's mapping can be obtained by substituting $s = j\,\omega$ in $G(j\omega)$. $H(j\omega)$. This will give the polar-plot of $G(j\omega)\,.\,H(j\omega)$ for ω from 0 to ∞.

Part II : Complete - j ω axis :

It's mapping is the mirror image of polar - plot of part I.

Part III : Right half of s-plane :

It's mapping is obtained by substituting $s = R_e^{j\theta}$; R → ∞; θ varies from + 90° to – 90°.

Part IV : Small semi-circular indent:

It's mapping is obtained by substituting $s = \varepsilon \cdot e^{j\theta}$, ε → 0. θ varies from — 90° to + 90° through 0°.

BODE PLOT

The plots of |G(jω)| in db and ∠ G(jω) in degrees verses frequency (ω) in rad/sec together are called as *Bode Plot.* For this purpose semi-log graph paper is used. Frequency ω is taken along log-scale (x-axis). Hence the scale is in multiples of 10.

If $\frac{\omega_1}{\omega_2} = 10$ it is called *decade* and if $\frac{\omega_2}{\omega_1} = 2$ it is called as *octave.*

For magnitude-plot along Y-axis (normal scale) 20 log |G(jω)| is taken. Its unit is decibel (db). For phase angle plot along Y-axis (normal scale) ∠ G(jω) in degrees is taken.

Advantages of Bode Plot

(i) As log of |G(jω)| is used in plotting the curve, the product factors of |G(jω)| are converted into additive terms.

(ii) In most of the system, the exact function plots can be represented approximately by straight line asymptotes. The accurate curves are then obtained by adding corrections to approximate plots.

(iii) The polar-plot and magnitude verses phase-angle-plots can be easily computed from the Bode-Plot.

(iv) The Bode-plot is constructed for open-loop T.F. From the Bode-Plot the idea about the stability of closed-loop system can be obtained.

(v) Low and high frequency regions can be brought into focus simultaneously.

Limitation

The closed-loop frequency response specifications can not be obtained directly from the Bode-Plot.

Bode-Plot of Standard Factors

(i) G*ain factor 'k'* : M = 20 log |k| = Constant (k' db)

∴ The magnitude plot will be a straight line parallel to X-axis; passing through k db on Y – axis.

$$\Phi = \angle G(j\omega) = \tan^{-1}\frac{(\text{Imaginary part})}{(\text{Real part})} = \tan^{-1}\left(\frac{0}{k}\right) = 0°$$

∴ Φ will remain constant i.e. 0°.

(ii) *Poles or Zeros at origin* $\left[\frac{1}{(j\omega)^r} \text{ or } (j\omega)^r\right]$:

Let us consider poles at origin i.e. $\frac{1}{(j\omega)^r}$

$$\therefore \quad M = 20\log\left[\frac{1}{(j\omega)^r}\right] = \{-(20\times r)\log\omega\}\,d\,b$$

Let $\omega_1 = 1$; ∴ M = 0 d b

and if $\omega_2 = 10$; ∴ M = (– 20 × r) d b

∴ The straight line will pass through points [1.0,d,b] and [10, – 20 r d b]

$$\therefore \text{ Slope } = \frac{y_2 - y_1}{x_2 - x_1} = \frac{-20r - 0}{\log 10 - \log 1} = -20 \text{ r db/dec.}$$

(As along x-axis it is log scale).

∴ Magnitude plot will be a straight line of slope – 20×rdb/dec.

Now

$$\therefore \Phi = \angle G(j\omega) = \tan^{-1}\left[\frac{\text{Imaginary part}}{\text{Real part}}\right]$$

$$\therefore \Phi = -\tan^{-1}\frac{(\omega)^r}{0} = (-90\times r) \text{ deg}$$

∴ Φ will remain constant i.e. (– 90×r) deg

Similarly for zeros : 'M' plot will be a straight line of slope + 20 × r db/dec and phase-angle will remain constant i.e. (90×r) deg.

(iii) *Poles/zeros on real axis* :

Let us consider a pole

say $\frac{1}{1 + j\omega T}$

$$M = 20\log\left|\frac{1}{1 + j\omega T}\right| = 20\log\left(\sqrt{1 + \omega^2 T^2}\right)$$

For constructing the Magnitude-plot for such terms the complete region is divide into two parts i.e.

(i) for $\omega < 1/T$ and

(ii) for $\omega > \frac{1}{T}$

The frequency $\omega = \frac{1}{T}$ is called as *corner frequency* $|\omega_c|$ The magnitude plot is constructed with following assumptions :

(i) For $\omega < \frac{1}{T}$; $1 + w^2 T^2 \approx 1$;

$\therefore M = 0$ db

(ii) For $\omega > \frac{1}{T}$; $1 + w^2 T^2 \approx w^2 T^2$;

$\therefore M = -20 \log \omega T$

$\therefore$ at $\omega = \frac{1}{T}$; $M = 0$ d b

and at $\omega = \frac{10}{T}$; $M = -20$ d b

$\therefore$ The magnitude-plot will be a straight line of slope – 20 d b/dec From the corner frequency and upto the corner frequency M = 0 d b.

$$\Phi = -\tan^{-1} \omega T ;$$

Hence for various values of ω, Φ can be computed and by joining all the points by a smooth curve, phase-angle plot can be obtained.

Note : at $\omega \to 0$; $\Phi \to 0°$; at $\omega = \omega_c = \frac{1}{T}$; $\Phi = -45°$ and at $\omega \to \infty$; $\omega \to -90°$.

If there are say r no. of poles at the same location i.e. for $\frac{1}{(1 + j\omega T)^r}$;

the magnitude plot will be a straight line of slope $-20 \times r$ d b/dec from the corner frequency and upto the corner frequency it will be 0 db; and

$$\Phi = -r \times \tan^{-1} \omega T.$$

Similarly for a zero; say $(1 + j\omega T)^r$; magnitude plot will be a straight line of slope $20 \times r$ db/dec from the corner frequency and upto the corner frequency it will be 0 db and $\Phi = (r \times \tan^{-1} \omega T)$ deg.

(iv) *Complex-poles and zeros :*

Let us consider complex poles ;

say $\left\{\frac{1}{(1-u^2)\pm 2j\zeta u}\right\}$

$M = -20 \log [(1 - u^2)^2 + 4 \zeta^2 u^2]^{1/2}.$

Here $u = \frac{\omega}{\omega_n}$

In this case ω_n is the corner frequency

i.e. When u = 1 [$\therefore \omega = \omega_c = \omega_n$]

$\therefore$ for u < 1; M $\approx$ 0 db

and for u > 1; M $\approx -20 \log \sqrt{(u^2)^2} = -20 \log u^2$.

= [– 40 log u] db

$\therefore$ at u = 1; M = 0 db and

at u = 10; M = – 40 db

$\therefore$ Magnitude plot will be a straight line of 0 db upto the corner frequency and then it will be a straight line of slope – 40 db/dec.

$$\Phi = -\tan^{-1}\left\{\frac{2\zeta u}{1-u^2}\right\};$$

hence for various values of ω; Φ is computed and then all the point are joined by a smooth curve.

Note :

As $u \to 0$, $\Phi \to 0°$ at u = 1 i.e. $\omega = \omega_n = \omega_c$ $\Phi = -90°$ and as $u \to \infty$; $\Phi \to -180°$.

Similarly for complex zeros i.e.

$\left\{(1-u^2)\pm j2\zeta u\right\}$; magnitude will be zero db upto the corner frequency (ω_c) and then it will be a straight line of slope 40 db/dec; and

$$\Phi = \tan^{-1}\left[\frac{2\zeta u}{1-u^2}\right]$$

Note : Let us consider a pole at origin i.e. $\frac{1}{j\omega}$

$\therefore$ M = – 20 log ω.

$\therefore$ For this when ω = 1;

M = 0 db and for ω = 10; M = – 20 db

$\therefore$ The plot will be a straight line of slope – 20 db/dec

Also when ω = 2; M = – 6 db

$\therefore$ Slope = – 6 db/octave

$\therefore$ $\omega_2 = 10\,\omega$, is called as one decade.

and $\omega_2 = 2\,\omega$, is called as one octave.

$\therefore$ Slope of – 20 db/dec = – 6 db octave

Corrections

In the previous section, it was observed that for gain (K) and poles/zeros at origin the magnitude plots are straight lines; whereas for poles/zeros on real axis as well as for complex poles/zeros we make the magnitude plots straight lines (changing the slope at corner frequency) by assumption. Hence we must know the corrections to be added.

(i) *Poles/zeros on the real-axis :*

Consider a pole i.e. $\dfrac{1}{1+j\omega T}$

The actual plot has magnitude; $M = -20\log\{1+\omega^2T^2\}^{1/2}$

$$= -10\log\{1+\omega^2T^2\} \text{ upto the corner frequency } \left(\omega_c = \frac{1}{T}\right),$$

We had assumed; M = – 20 log (1)

$\therefore$ Upto the ω_c.

Δ_1=Error=– 10 log {1 + ωsup T^2} – (– 20 log 1)

$\therefore \quad \Delta_1 = -10\log\{1+\omega^2T^2\}$

$\therefore \quad [\Delta_1]_{\frac{1}{T}}$ = Error at ω_c.

$$= -10\log\{1+\omega^2T^2\}\big|_{\omega=\omega c=1/T} = -3 \text{ db}$$

$[\Delta_1]_{\frac{1}{2T}}$ = Error One octave below ω_c.

$$= -10\log\{1+\omega^2T^2\}\big|_{\omega=1/2T} = -1 \text{ db}$$

Similarly the error Δ_2 for frequency $\omega > \omega_c$ will be given by;

$\Delta_2 = -10\log\{1+\omega^2T^2\} - \{-20\log\omega T\}$;

as for $\omega > \omega_c$; $M = -20\log\omega T$

$\therefore \Delta_2 = -10\log\{1+\omega^2T^2\} + 20\log\omega T$

$\therefore\ [\Delta_2]_{\frac{2}{T}}$ = Error one octave above the ω_C.

$$= -10\log\left\{1+\left(\frac{2}{T}\right)^2 T^2\right\}+20\log\left(\frac{2}{T}.T\right) = -1 \text{ db}$$

Similarly the corrections at other ferquencies are also evaluated. The graph of Δ_1 and Δ_2 verses log ω.

(ii) *Complex poles/zeros :*

Let us consider poles

Actual magnitude;

$M = -10 \log \{(1-u^2)^2 + 4\zeta^2u^2\}$

$\Delta_1 = -10 \log \{(1-u^2)^2 + 4\zeta^2u^2\}$

As upto u = 1; $M \approx -20 \log 1 = 0$

and $\Delta_2 = -10 \log \{(1-u^2)^2 + 4\zeta^2u^2\} + 40 \log u$

As after u = 1; $M \approx -40 \log u$

General Procedure for Constructing the Bode-plot

Step 1 : Bring G(jω) . H(jω) in standard form.

Step 2 : Write down each actor in a table, write remarks.

Step 3 : Draw the magnitude plot for each factor clearly.

Step 4 : Add all points clearly and get the resultant curve.

Step 5 : Draw the phase plot for each factor clearly.

Step 6 : Add all points clearly and get the resultant curve.

Stability Checking

Following terms are also associated with the Bode-Plot and they are useful for stability checking of closed-loop system. Note that we are constructing the Bode-plot for open-loop T.F. G(jω) . H(jω).

(i) *Gain cross-over frequency (ω_g) :* It is the frequency at which magnitude plot crosses 0 db line i.e. it is the frequency where M = 0 db.

(ii) *Phase cross-over frequency (ω_p) :* It is the frequency at which phase-angle plot crosses – 180° axis. i.e it is the frequency where Φ = – 180°

(iii) *Gain-Margin (GM) :* GM = – {Magnitude of G(jω).H(jω) at ω_p in db}

(iv) *Phase-Margin (Φ_{PM}) :* (Φ_{PM}) = 180° + Angle of G(jω) . H(jω) at ω_g.

Note : GM and Φ_{PM} are concepts of open-loop stable systems. The corresponding closed-loop system will be stable if GM and Φ_{PM} are positive; i.e. if $\omega_g < \omega_p$.

Gain Adjustment for required GM and Φ_{PM}

Let us consider a stable closed-loop system. Hence for this system GM and Φ_{PM} will be positive and $\omega_g < \omega_p$. Now if we change the gain 'K' of the system, the magnitude plot will change but its phase-angle plot will remain unchanged. Also as k is increased ω_g will also increase, hence GM and Φ_{PM} will reduce. When $\omega_g = \omega_p$; GM = Φ_{PM} = 0 and system will become oscillatory.

Let (GM) = Gain – margin of the system and (GM)' be the specified GM. Let k is the gain corresponding to GM and k' is the required gain for achieving (GM)'.

At $\omega = 1$; gain of the system = 20 log k = M; say

Let Δ GM = (GM) – (GM)'

$\therefore$ at $\omega = 1$; required gain = M + Δ GM

$\therefore$ 20 log k' = M + Δ GM ...(1)

$\therefore$ The required gain for the specified gain-margin can be obtained using equation (1).

Similarly; Let Φ'_{PM} = specified P.M. and Φ' = Angle at this P.M.

$\therefore \Phi' = -180 + \Phi'_{PM}$.

From the Bode-plot find the gain at this phase-angle. Let it be Δ M.

$\therefore$ 20 log k' = M – Δ M ...(2)

Hence the required gain k' for the specified P.M. can be computed using equation (2).

Determination of maximum gain (k_{max}) for stability of the system

$\therefore$ 20 log k_{max} = M + GM ..(3)

DETERMINATION OF T.F. FROM THE BODE-PLOT

In the process of recovery of transfer functions, two possibilities exist. Either asymptotic curves are given or experimental data is given. In the latter, correction needs to be applied.

Transfer Function from Asymptotic Plots

In the recovery from asymptotic plots,low frequency region slopes can be 0 db i.e. horizontal lines or multiples of 20 db/dec. Here two cases may be created.

Case 1 : Low frequency region, slope is 0 db i.e. horizontal line.

Case 2 : Low frequency region slope is a multiple of 20 db/dec.

(A) **Recovery of Transfer Function for Case 1 and Case 2**

Case 1

Step 1 : To find value of k, y intercept in db must be given or found out. Say it is T db, then

$$20 \log k = T \quad ...(1)$$

From (1) k can be found out.

If T is not given it can be traced by finding from slopes and intercepts for the curve.

Step 2 : Remove the value of k i.e. T db. This shifts the curve to 0 db for low frequency part. Now follow same steps for recovery of terms as in Case 2 i.e. from step 5 onwards.

Case 2

Step 3 : The first term for this case is always a zero or pole at origin of order depending on the value of slope i.e. – 20 db means $1/j\omega$. – 40 db means $\frac{1}{(j\omega)^2}$ etc. and 20 db means $j\omega$, 40 db means $(j\omega)^2$ etc.

Step 4 : The value at $\omega = 1$ is say T_2db

i.e. $20 \log k = T_2$

$\therefore$ K can be found out.

Step 1 and Step 2 are unique for case 1. Step 3 and 4 for case 2, Step 5 onwards is common.

(B) **Recovery of other terms**

The following steps are to be repeated (step 5 to step 8)

Step 5 : Where is the change of slope taking place (say at a)

Step 6 : What is the change in slope ?

{Change in slope = new slope – old slope}

Step 7 : First order terms have slope ± 20 db/dec. second order terms have slope 40 db/dec and so on. In general n × 20 db/dec say.

Step 8 : The term is

(1) $\left(1 + j\frac{\omega}{a}\right)^n$ if slope is positive and

(2) $\dfrac{1}{\left(1+j\dfrac{\omega}{a}\right)^n}$ if slope is negative.

Nichol's Chart

This is very useful chart for determining the closed-loop frequency response for the specified open loop T.F. By constructing the Bode-plot and Polar-plot we can only determine the GM, PM, ω_q & ω_p; but we can not find the closed-loop frequency response specifications i.e. M_r, ω_r, ω_b etc. Nichol's chart not only gives these specifications but also gives the idea about gain adjustment for the desired frequency response specifications.

Nichol's chart consist of constant 'M' & 'N' circles. It has phase-angle in degrees along X-axis and magnitude in db along Y-axis. For the given open-loop T.F. we determine magnitude of db & phase-angle in degrees for various values of frequencies. Then we plot gain-phase plot on Nichol's chart. After drawing the gain-phase plot we can determine following from the Nichol's chart.

(i) Gain-margin, Phase-margin, Gain cross-over frequency and Phase cross-over frequency.

(ii) Stability of the corresponding closed loop system.

(iii) Closed-loop frequency response

(iv) Closed-loop frequency response specifications i.e. resonant peak, resonance frequency and band-width.

Gain for specified closed loop frequency response specifications, as well as for specified GM and PM.

Frequency Response (Frequency Domain) Analysis

The magnitude and phase-relation between the sinusoidal input and the steady state output is termed as *frequency response* i.e. the steady-state response of a system to the sinusoidal input is termed as frequency response.

For frequency response analysis ; input is

$r(t) = A \sin \omega t$

Where A = Amplitude of the sinusoidal input and

ω = Frequency in rad/sec.

For this input the output is in the form.

$\therefore c(t) = B \sin (\omega t + \Phi)$

Where B = Amplitude of the output and

Φ = Phase-shift w.r.t. input

Conclusions :

(i) The output is also sinusoidal and has the same frequency as that of the input.

(ii) The output has different amplitude and has a phase-difference of Φ with the input.

(iii) The output is independent of amplitude of the input signal.

(iv) The system only changes the amplitude and phase of the output w.r.t. the input.

The frequency response test is carried out by keeping 'A' fixed and determining B and Φ for a suitable range of frequencies.

Methods used in Frequency-Response Analysis

The frequency response characteristic has two plots :

(i) Magnitude function plot. (ii) Phase function plot.

The variable for both these plots is ω i.e. the angular frequency. Following methods are used for analysis :

(i) Polor plot (ii) Bode plot

(iii) Magnitude verses phase-angle plot.

Advantages of Frequency Domain Analysis

(i) Frequency response test can be easily carried out in a laboratory with good accuracy. This is useful if transfer function is difficult to be obtained by analytical techniques.

(ii) Design of open-loop T.F. for specified closed loop performance is easier in frequency domain.

(iii) Effects of parameter variations and noise disturbance are easy to visualize in frequency response.

(iv) The performance specifications and response of a system in frequency domain can be related with it's performance specifications and response in time domain.

FREQUENCY-DOMAIN SPECIFICATIONS

For a closed-loop system shown we have;

Closed loop T.F. $T(s) = \dfrac{C(s)}{R(s)} = \dfrac{G(s)}{1+G(s).H(s)}$

For writing the sinusoidal T.F.; s is replaced by jω (as it is for steady state only)

$$\therefore \text{ Sinusoidal T.F. } = T(j\omega) = \frac{C(j\omega)}{R(j\omega)} = \frac{G(j\omega)}{1+G(j\omega)H(j\omega)}$$

The different specifications in frequency domain are :

(i) *Resonant Peak or Peak Resonance (M_r) :* This is the maximum value of the magnitude.

$\therefore M_r$ = Maximum value of $|T(j\omega)|$ as ω is varied.

(ii) *Resonance frequency (ω_r) :* It is the frequency at which peak resonance occurs.

(iii) *Band-width (ω_b) :* It is the range of frequencies over which $|T(j\omega)|$ is equal to or greater than 70.7% of its zero frequency level.

(iv) *Cut-off frequency (ω_{cut}) :* It is the frequency at which $|T(j\omega)|$ is 70.7%

(v) *Gain cross-over frequency (ω_g) :* It is the frequency at which $|G(j\omega).H(j\omega)| = 1$ i.e. 20log M = zero decibel (db).

(vi) *Phase cross-over frequency (ω_p) :* It is the frequency at which.

$\angle G(j\omega) . H(j\omega) = 180^o$.

(vii) *Phase-margin (Φ_{PM}) :* $\Phi_{PM} = 180^o + \angle G(j\omega) . H(j\omega) |_{\omega = \omega g}$.

(viii) *Gain-Margin (GM) :*

$$GM = \frac{1}{|G(j\omega).H(j\omega)|_{\omega = \omega p}}$$

Note : M_r, ω_ρ, ω_b and ω_{cut} are the specifications of closed loop frequency response i.e. they are defined for T(jω). Whereas ω_q, ω_ρ, Φ_{PM} and GM concepts are applicable to the open-loop stable systems only i.e. they are defined for G(jω) . H(jω).

CORRELATION BETWEEN TIME AND FREQUENCY RESPONSE SPECIFICATIONS

For a standard 2nd order unity feedback system

$$T(s) = \frac{C(s)}{R(s)} = \frac{\omega_n^2}{s^2 + 2\zeta\omega_n s + \omega_n^2}$$

$$\therefore T(j\omega) = \frac{C(j\omega)}{R(j\omega)} = \frac{\omega_n^2}{(j\omega)^2 + 2\zeta\omega_n (j\omega) + \omega_n^2}$$

$$= \frac{\omega_n^2}{\omega_n^2\left\{-\left(\frac{\omega}{\omega_n}\right)^2 + 2\zeta j\left(\frac{\omega}{\omega_n}\right) + 1\right\}}$$

$$T(j\omega) = \frac{1}{1 + j2\zeta\left(\frac{\omega}{\omega_n}\right) - \left(\frac{\omega}{\omega_n}\right)^2}$$

Let $\frac{\omega}{\omega_n} = u \rightarrow$ Normalized driving signal frequency.

$$\therefore \quad T(ju) = \frac{1}{1 + j2\zeta u - u^2} \qquad \text{...(1)}$$

$$\therefore \quad M = |T(ju)| = \frac{1}{\left\{(1-u^2)^2 + (2\zeta u)^2\right\}^{\frac{1}{2}}}$$

and $\Phi = \angle T(ju) = -\tan^{-1}\left(\frac{2\zeta u}{1-u^2}\right)$ ...(2)

$\therefore$ For a input of $r(t) = A \sin \omega t$, output will be $c(t) = M \sin(\omega t + \Phi)$ i.e.

$$c(t) = \frac{1}{\left\{(1-u^2)^2 + (2\zeta u)^2\right\}^{\frac{1}{2}}} \sin\left\{\omega t - \tan^{-1}\left(\frac{2\zeta u}{1-u^2}\right)\right\} \qquad \text{...(3)}$$

Resonant Peak and Resonance Frequency

According to the definition, resonant peak (M_r) is the maximum value of $|T(j\omega)|$

Now $|T(ju)| = \frac{1}{\left\{(1-u^2)^2 + (2\zeta u)^2\right\}^{\frac{1}{2}}}$

$$\therefore \quad \frac{d|T(ju)|}{du} = \frac{4u - 8u\zeta^2 - 4u^3}{\left\{2(1-u^2)^2 + (2\zeta u)^2\right\}^{\frac{3}{2}}}$$

For maximum $|T(ju)|$; $\frac{d|T(ju)|}{du} = 0$

This gives $4u - 8u\,\zeta^2 - 4u^3 = 0$

i.e. $4u\,[1 - 2\zeta^2 - u^2] = 0$

$\therefore$ $u = 0$ or $u = \sqrt{1-2\zeta^2}$

But $u = 0$ is not possible.

$\therefore$ at $u = \sqrt{1-2\zeta^2}$; $|T(j\omega)|$ will be maximum ...(4)

$\therefore$ This will be u_r i.e. normalised resonance frequency.

$$\therefore\ u_r = \frac{\omega_r}{\omega_n} = \sqrt{1-2\zeta^2} \qquad ...(5)$$

$$\therefore\ \omega_r = \omega_n \cdot \sqrt{1-2\zeta^2}$$

Substitute value of u_r from eq. (4) in (1) to get M_r.

$$\therefore\ M_r = \frac{1}{\left\{4\zeta^4 + 4\zeta^2(1-2\zeta^2)\right\}^{1/2}}$$

$$\therefore\ M_r = \frac{1}{2\zeta\sqrt{1-\zeta^2}}$$

Again put the value of u_r from eq. (4) into (1) to get Φ_r; then

$$\therefore\ \Phi_r = -\tan^{-1}\left\{\frac{\sqrt{1-2\zeta^2}}{\zeta}\right\} \qquad ...(6)$$

From (5), (6) and (7) following observations can be made.

(1) As $\zeta \to 0$; $\omega_r \to \omega_n$, $M_r \to \infty$ and $\Phi_r \to -90^\circ$.

(2) If $\sqrt{1-2\zeta^2} = 0$ i.e. $\zeta = 0.707$ then $\omega_r = 0$;

$M_r = 1$ and $\Phi_r = 0^\circ$.

(3) For $0 < \zeta < 0.707$, $\omega_r < \omega_n$, $M_r > 1$ and $\Phi_r < 0^\circ$.

(4) For $\zeta > 0.707$; ω_r becomes imaginary. Hence the equation is valid for $\zeta > 0.707$ only. Also for $\zeta > 0.707$; $M_r < 1$, there will be no resonant peak.

Band-width (ω_b)

Cut-off frequency (ω_c) is that point where $|T(j\omega)|$ is 0.707.

Usually control systems are low pass filters; hence

$|T(j\omega)| = 1$ for zero frequency due to which band width (B.W.) becomes equal to ω_c.

$\therefore$ at $\omega = \omega_c = \omega_b$; $|T(jw)| = 0.707$.

Put this into equation (1) above

$$\therefore \; 0.707 = \frac{1}{\left\{(1-u_b^2)^2 + (2\,ub\zeta)^2\right\}^{\frac{1}{2}}}$$

$$ub = \frac{\omega_b}{\omega_n}$$

$$\therefore \; ub = \left\{1 - 2\zeta^2 + \sqrt{2 - 4\zeta^2 + 4\zeta^4}\right\}^{\frac{1}{2}}$$

$$\therefore \; \omega_b = \omega_n \left[1 - 2\zeta^2 + \sqrt{2 - 4\zeta^2 + 4\zeta^4}\right]^{1/2} \qquad ...(7)$$

Eq. (5) to (7) show relationships between frequency response specifications and time response specifications.

SIGNIFICANCE OF THE TERMS USED IN FREQUENCY-DOMAIN

(i) *Resonant Peak (M_r) :*

$$M_r = \frac{1}{2\zeta\sqrt{1-\zeta^2}}$$

This factor is indicative of damping (ζ) of the system. Therefore it indicates relative stability of the closed-loop system. Larger value of M_r indicates larger overshoots (M_p). Acceptable value of M_r is in between 1.1 and 1.5.

(ii) *Resonance Frequency (ω_r) :*

$$\omega_r = \omega_n \sqrt{1 - 2\zeta^2}$$

This factor is indicative of natural frequency (ω_n). Therefore it indicates speed of response.

(iii) *Band-width (ω_b) :*

$$\omega_b = \omega_n \left[1 - 2\zeta^2 + \sqrt{2 - 4\zeta^2 + 4\zeta^4}\right]^{1/2}$$

A large BW means higher frequencies are easily passed through the system; hence it indicates faster rise time. A small BW means only few frequencies are passed; hence it indicates sluggish response.

Therefore it gives idea about transient-response of the system.

(iv) *Phase-margin (Φ_{PM}) :*

It is a measure of relative-stability of the system. If phase-margin is positive then the corresponding closed loop system is stable. Φ_{PM} indicates how much the system angle can be increased to cause the system to become unstable from a stable condition. Φ_{PM} of 30° to 35° results in a relatively good degree of relative stability.

(v) *Gain-Margin (GM) :*

It is also indicative of relative-stability of the system. If GM is positive then the corresponding closed-loop system is stable. GM indicates how much gain can be increased to cause system instability. GM of about 6 db results in good relative-stability.

Note : GM and Φ_{PM} are concepts of open -loop stable system only.

EFFECT OF ADDING A ZERO AND A POLE

When a zero is added in the forward-path of a second order system :

(i) BW increases; hence rise time reduces.

(ii) Settling time increases.

(iii) System becomes more stable.

When a pole is added :

(i) BW decreases; hence rise time increases.

(ii) M_r increases.

(iii) System becomes less stable.

POLAR PLOTS

Defintion : The locus of all points of $|G(j\omega)| \angle G(j\omega)$ i.e. $G(j\omega)$ as ω is varied from zero to infinity is called as polar plot; i.e. If $|G(j\omega)|$ and $\angle G(j\omega)$ are plotted together as a function of ω, the result is polar plot.

Advantages of Polar Plot

(i) It indicates the frequency response characteristic over the entire range of frequencies in a single plot.

(ii) The stability of the system can be investigated using polar plot.

Limitations of Polar Plot

(i) Closed-loop frequency domain specifications (M_r, ω_r and ω_b) can not be obtained from the polar plot.

(ii) Addition of poles and zeros to an existing system requires the complete recalculation of frequency response.

Note : For analysis purpose only the general shape of the polar plot is needed. Also only a few regions and points are of interest. Hence in the polar plot the points at $\omega = 0$ and $\omega = \infty$ and intersection point with X and Y - axis are of importance and they give the general shape of the polar plot.

General Procedure For Constructing the Polar Plots

Step 1 : Determine the T.F. of the system in frequency domain i.e. $G(j\omega)$

Step 2 : Write the expression for magnitude i.e. $|G(j\omega)|$ and phase angle i.e $\angle G(j\omega)$

Step 3 : Obtain $|G(j\omega)|$ and $\angle G(j\omega)$ at $\omega = 0$ and at $\omega = \infty$. These points will give the starting and ending points of the polar plot. These will be in polar frame of reference. They may be converted into rectangular co-ordinates also.

Step 4 : For computing the intersection points with X and Y axis any of the following techniques (a or b) may be used :

(i) Substitute $\angle G(j\omega)$ = Angle of the axis for which intersection is being considered in the equation. Hence determine the frequency i.e. ω.

(ii) Substitute this frequency in the equation of $|G(j\omega)|$. This will give the intersection point.

Same procedure can be used for finding intersection point with any angle - axis.

(i) Express $|G(j\omega)|$ in real and imaginary parts.

(ii) If intersection with real axis is required; then $\{G(j\omega)\} = 0$.

$\therefore$ Compute the frequency ω.

(iii) Substitute this frequency in the equation of real $\{G(j\omega)\}$. This will give the intersection point.

Same procedure is used for finding intersection point with real axis.

Step 5 : Locate few more points; if required; and sketch the Polar-Plot.

DETERMINATION OF GM AND Φ_{PM} FROM POLAR-PLOTS

Let us assume that for a certain system, the polar. The system must be a open-loop stable system. Now as per the definitions of Gain Margin, it is the factor by which gain can be increased to drive the system to the verge of instability, and mathematically;

$$GM = \frac{1}{|G(j\omega)|_{\omega = \omega_p}}$$

Where ω_p is the phase cross-over frequency

i.e. the frequency at which $\angle G(j\omega) = -180^\circ$.

Similarly, Phase margin $\Phi_{PM} = 180 + \angle G(j\omega)\,|_{\omega = \omega g}$.

Where ω_g is the gain cross-over frequency i.e. the frequency at which $|G(j\omega)| = 1$.

Graphical Method :

(1) Construct the polar-plot of the given system.

(2) Find the frequency at which the plot intersects -180°. This will be ω_p.

(3) Find the intersection point; say 'a'. This will be $|G(j\omega)|$ at ω_g.

$$\therefore \ \text{GM} = \frac{1}{a} \text{ or GM} = 20 \log\left(\frac{1}{a}\right) \text{db}$$

(4) With centre at origin and radius = 1,draw a circle which intersects the polar-plot; say at 'b'. ∴ dist ob = 1

The frequency at this point will be ω_g.

(5) $\Phi_{PM} = \angle$ ob w.r.t. -180° axis $= \theta^\circ$.

Analytical Method :

(i) Equate $\angle G(j\omega)$ to 180° and find the frequency. This will be ω_p.

(2) Find $|G(j\omega)|$ by substituting $\omega = \omega_p$.

(3) $$\text{GM} = \frac{1}{|G(j\omega)|_{\omega = \omega p}}$$

(4) Equate $|G(j\omega)|$ to 1 and find the frequency. This will be ω_g.

(5) $\Phi_{PM} = 180^\circ + \angle G(j\omega)\,|_{\omega = \omega g}$.

GAIN ADJUSTMENT FOR SPECIFIED Φ_{PM} AND GM

Let us assume that the polar-plot of a particular system is as shown for k = 1

$$\therefore \ \text{GM} = \frac{1}{a} \text{ and PM} = \theta^\circ. \qquad ...(1)$$

Let $(\text{GM})_{sp}$ = Specified GM

The graph must intersect at point say 'a' for this specified GM such that

$$[\text{GM}]_{sp} = \frac{1}{a'} \qquad ...(2)$$

If $[\text{GM}]_{sp} > \text{GM}$; then $a' < a$

and If $[GM]_{sp} < GM$; then $a' > a$

For $k = 1$; $|G(j\omega)|_{\omega = \omega g} = a$

Let k' be the gain to give the specified GM.

$\therefore k' |G(j\omega)|_{\omega = \omega g} = a'$. $\therefore \dfrac{k'}{k} = \dfrac{a'}{a}$

$$\therefore k' = \frac{a'}{a} \qquad ...(3)$$

$\therefore$ In order to increase the GM; gain should be reduced. Let Φ_{PM} = specified Φ_{PM} and k' = corresponding gain.

$\therefore$ The unit circle must interact the polar plot at the point so that line joining this point with origin makes angle $\theta' = \Phi'_{PM}$ w.r.t. -180°.

$$\therefore k' = \frac{ob'}{ob} \qquad ...(4)$$

Analytical Method :

Step 1 : Compute ω_p for $k = 1$

Step 2 : $\dfrac{1}{|G(j\omega)|_{\substack{\omega = \omega p \\ k = k'}}} = [GM]_{sp}$

$\therefore$ Compute k'.

$$\text{i.e. } k' |G(j\omega)|_{\omega = \omega p} = \frac{1}{[GM]_{sp}} \qquad ...(5)$$

Step 3 : Let ω'_g be the new gain cross-over frequency (for k').

$$\therefore \angle G(j\omega)|_{\omega'g} = \Phi_{sp} - 180^o. \qquad ...(6)$$

$\therefore$ Compute ω'_g.

Step 4 : As per definition of gain cross-over frequency, at this frequency $|G(j\omega)| = 1$.

$\therefore k' |G(j\omega)|_{w = wg} = 1$ $\qquad \therefore$ Compute k'. $\qquad$ ...(7)

Note :

(i) GM and Φ_{PM} are the concepts of the open-loop stable system only.

(ii) If $\omega_g < \omega_p$; the corresponding closed-loop system is stable.

(iii) If GM (in db) and Φ_{PM} are positive; then corresponding closed-loop system is stable.

(iv) As k (gain) of the system goes on increasing, the GM and Φ_{PM} goes on reducing; hence relative stability of the system also goes on reducing.

DETERMINATION OF MAXIMUM GAIN (KMAX) TO KEEP SYSTEM STABLE

Let 'a' be the intersection of polar-plot and – 180° axis for k=1.

$$\therefore \text{ GM} = 20 \log \left[\frac{1}{a}\right] \text{ db.}$$

For stable system a < 1; hence GM will be positive.

As k goes on increasing 'a' shifts towards left hand side and at a = 1 the system has GM = 0 db.

Let a = 1 corresponds to k_{max}.

$$\therefore k_{max} |G(j\omega)|_{\omega = \omega p} = 1 \quad \text{Also } |G(j\omega)|_{\substack{\omega = \omega p \\ k = 1}} = a$$

$$\therefore k_{max} = \frac{1}{a} \qquad \text{...(1)}$$

$$\text{i.e. } k_{max} = \frac{1}{|G(j\omega)|_{\substack{\omega = \omega p \\ k = 1}}} \qquad \text{...(2)}$$

SOLVED EXAMPLES

Example 1:

Draw the Bode plot for the following open loop transfer function. Indicate the gain and phase margin.

$$G(s) = \frac{2500\,(s+10)}{s(s+2)\,(s^2+30s+2500)}$$

Solution:

$$G(s) = \frac{2500\,(s+10)}{s(s+2)\,(s^2+30s+2500)}$$

$$= \frac{5\left(\frac{s}{10}+1\right)}{s\left(1+\frac{s}{2}\right)\left[\left(\frac{s}{50}\right)^2+\left(\frac{30}{50}\right)s+1\right]}$$

$$G(j\omega) = \frac{5\left(1+j\frac{\omega}{10}\right)}{j\omega\left\{1+j\frac{\omega}{2}\right\}\left\{\left(\frac{j\omega}{50}\right)^2+\left(\frac{30}{50}\right)j\omega+1\right\}}$$

Step 1 : Draw the magnitude plot. (See figure).

Step 2 :

(1) (20 + 14) = 34 db at $\omega = 0.1$ and slope of – 20 db/dec for $\omega < 2$.

(2) Slope of – 40 db/dec for $2 < \omega < 10$.

(3) Slope of = – 20 db/dec for $10 < \omega < 50$

(4) Slope of – 60 db/dec for $50 < \omega$.

Step 3 : Draw the phase plot (See previous figure)

Step 4 :

(1) $\theta = -90^\circ$ for $\omega < 0.4$

(2) Decreases linearly due to (4) $0.4 < \omega < 2$

(3) Remains constant for $2 < \omega < 10$

(4) Falls at same rate as (2) for $10 < \omega < 50$

(5) Falls at twice slopes as (4) for $50 < \omega < 250$

(6) Remains constant at approximate -270°, for $250 < \omega$.

Step 5 : $\omega_{gc} = 3$

Step 6 : PM = 48.6° as $(180 + (-131.4) = 48.6^\circ)$

Step 7 : $\omega_{pc} = 50$

Step 8 : GM = 36 db (36 db below 0 db line)

Since gain margin (GM) and phase margin (PM) are positive and system is stable.

Example 2:

Explain the Bode-Plot of unity feedback system having T.F.

$G(s) = e^{-j\omega T}$

Solution:

$$G(j\omega) = e^{-j\omega T} = \cos\omega T - j\sin\omega T$$

$$\therefore\ M = 20\log\{\cos^2\omega T + \sin^2\omega T\}^{1/2} = 0 \text{ db}$$

$$\therefore\ \Phi = -\tan^{-1}\left\{\frac{\sin\omega T}{\cos\omega T}\right\}$$

$$= -\tan^{-1}(\tan\omega T) = -\omega T \text{ rad}$$

$$\therefore\ \Phi = -\left\{\omega T \times \frac{180}{\pi}\right\} \text{ degree}$$

∴ The magnitude-plot of this term will be a straight line of 0 db magnitude and phase-angle will change with ω according to the given equation.

The term $e^{-j\omega T}$ is called as *transporta-tion lag* term. When it is used with any open-loop T.F.; [say $e^{-j\omega T}$ G(jω)]; it does not change the magnitude-plot of G(jω); but changes the phase-angle plot of G(jω). The phase-angle plot becomes more negative; hence ω_p reduces. As ω_p reduces GM and Φ_{PM} also reduce and hence relative stability of the system also reduces. If time-constant (T) is large then it may lead to unstable system operation.

Example 3:

For the system compute the gain of the system to ful-fill the following requirements.

(i) GM = 10 db (ii) GM = 40 db

(iii) PM = 40° (iv) Maximum gain for stability of the system.

Solution:

For the given system;

$G(s) = \dfrac{k}{s(s+1)(s+5)}$ [k = 1] the Bode plot is as constructed and we have obtained the following results :

ω_g = 0.2 rad/sec; ω_p = 2 rad/sec; GM = 28 db and Φ_{PM} = 90°. As GM and Φ_{PM} both are positive [$\omega_g < \omega_p$]; system is stable. Using equation (1), (2) and (3); this problem will be solved.

Note that magnitude of the given T.F. for k = 1 at ω = 1 rad/sec = M = 0 db.

(i) (GM) = Specified GM = 10 db

$\therefore \Delta GM = GM - (GM)' = 28 - 10 = 18$ db

Let k_1 be the required gain.

$\therefore 20 \log k_1 = M + \Delta GM = 0 + 18 = 18$ db

$\therefore k_1 = 8$

(ii) (GM)' = 40 db

$\therefore \Delta GM = 28 - 40 = -12$ db

Let k_2 be the required gain

$\therefore 20 \log k_2 = 0 - 12 = -12$ db

$\therefore k_2 = 0.25$

(iii) Φ'_{PM} = Specified P.M. = 40°

$\therefore \Phi' = -180^o + \Phi'_{PM} = -180^o + 40^o = -140^o.$

From the Bode Plot the gain say ΔM,

when $\angle G(j\omega) = -140^o$ is -14 db

Let k_3 be the required gain.

$\therefore 20 \log k_3 = M - \Delta M = 0 - (-14) = 14$ db

$\therefore$ $k_3 = 5$

(iv) Let k_{max} be the gain

$\therefore 20 \log k_{max} = M + GM = 0+28 = 28$ db

$\therefore \quad k_{max} = 25$

Example 4:

Determine the T.F. of the system whose Bode-Plot (magnitude only) is as given in Figure below.

Solution:

As low frequency region slope is 0 db,

This is case 1 problem

Step 1 : Y-axis intercept is not given. Hence to find it out. Intercept at 0 db axis is at $\omega = 4$.

Between $\omega = 3$ and 4 slope is -12 db/oct i.e. -40 db/dec.

[Since $\omega = 3$ and $\omega = 4$ are decade apart.

Note : Here plotting is corresponding to log ω hence $\omega = 3$ position means frequency 1000 and $\omega = 4$ means frequency 10^4 which are decade apart.]

Value of db at $\omega = 3$ is 40 db above $\omega = 4$ as line slopes 40 db is 1 decade

Value at $(\omega = 3)$ = values at $(\omega = 4) + 40$ db

$= 0 + 40 = 40$ db

Similarly $\omega = 2$ and $\omega = 3$ positions are 1 decade apart, slope is 20 db/dec.

$$\left[\text{Slope } 6\frac{\text{db}}{\text{oct}} = 20\frac{\text{db}}{\text{oct}}\right]$$

$\therefore$ value at $\omega = 2$ is

Value at $(\omega = 2)$ = Value $(\omega = 3) + 20$

$= 40 + 20 = 60$ db

Hence y-intercept is known.

$20 \log k = 60 \quad \therefore \log k = 3 \quad \therefore k = 1000$

Step 2 : The removal of k gives following curve shown.

Step 3 and 4 : Not needed.

Step 5 : 0 db line continues upto ω = 2 position for above fig.

i.e. frequency = 10^2 = 100 rad/sec

Change of slope is at 100 rad/sec.

∴ a = 100

Step 6 : Change in slope = new slope – old slope

= – 20 – (0) = – 20 db

Step 7 : n = 1

Step 8 : Term is $\dfrac{1}{\left(1+\dfrac{j\omega}{100}\right)}$.

Step 5 to 8 is repeated for ω = 3 position i.e. 10^3 rad/sec. The results are

a = 10^3,– 20 db/dec, n = 1

The term is $\dfrac{1}{\left(1+j\dfrac{\omega}{10^3}\right)}$. No further change in slopes.

∴ The transfer function is

$$T(j\omega) = \frac{1000}{\left(1+\dfrac{j\omega}{10^2}\right)\left(1+j\dfrac{\omega}{10^3}\right)}.$$

Example 5:

Sketch the polar-plot of

$$G(s) = \frac{6}{(s+1)\,(s+2)}$$

Solution:

$$G(s) = \frac{6}{(1+j\omega)\,(2+j\omega)}$$

$$\therefore\ |G(j\omega)| = \frac{6}{\sqrt{1+\omega^2}\,\sqrt{(2^2+\omega^2)}}$$

$$\angle G(j\omega) = -\tan^{-1}\omega - \tan^{-1}\frac{\omega}{2}$$

ω	$\lvert G(j\omega)\rvert$	$\angle G(j\omega)$	
0	3	0	→starting pt.
∞	0	– 180°	→ending pt.
$\sqrt{2}$	$\sqrt{2}$	– 90°	
2	0.95	– 108.44°.	

As the point is starting from 0° and ending at – 180°; it will pass through – 90° axis.

∴ let us find out the intersection point. [use step 4-a]

$$\angle G(j\omega) = -\tan^{-1}\omega - \tan^{-1}\frac{\omega}{2}$$

$$\therefore -90^\circ = -\tan^{-1}\omega - \tan^{-1}\frac{\omega}{2}$$

Solving this equation we get $\omega = \sqrt{2}$

$$\therefore \ |G(j\omega)|_{\omega=\sqrt{2}} = \frac{6}{\sqrt{1+\sqrt{2}^2}\,.\sqrt{2^2+\sqrt{2}^2}} = \sqrt{2}$$

∴ The polar-plot will be as shown in figure below.

Example 6:

Sketch the Polar-plot for

$$G(s) = \frac{10}{s(s+1)(s+2)}.$$

Solution:

$$G(j\omega) = \frac{10}{j\omega(1+j\omega)(2+j\omega)}$$

$$\therefore \ |G(j\omega)| = \frac{10}{\omega\sqrt{1+\omega^2}\,\sqrt{4+\omega^2}}$$

$$\angle G(j\omega) = -90^\circ - \tan^{-1}(\omega) - \tan^{-1}\frac{\omega}{2}$$

ω	$\lvert G(j\omega)\rvert$	$\angle G(j\omega)$	
0	∞	– 90°	→starting pt.
∞	0	– 270°	→ending pt.
$\sqrt{2}$	1.77	– 180°	→Crossing with ± 180° axis.

As the point is starting at – 90° and terminating at – 270°; it will cross ± 180°. Hence the interaction point can be obtained by step (4-a) of procedure

$\angle G(j\omega) = -180^\circ = 90^\circ - \tan^{-1}\omega - \tan^{-1}(\omega/2)$

$\therefore \tan^{-1}\omega + \tan^{-1} 0.5\,\omega = 90^\circ$

$\therefore \tan\{\tan^{-1}\omega + \tan^{-1} 0.5\omega\} = \tan 90^\circ = \infty.$

$$\therefore \frac{\omega + 0.5\,\omega}{1 - 0.5\,\omega^2} = \infty$$

$\therefore 1 - 0.5\,\omega^2 = 0$

$\therefore \omega = \sqrt{2}$

$$\therefore |G(j\omega)|_{\omega = \sqrt{2}} = \frac{10}{\sqrt{2} \times \sqrt{1+2} \times \sqrt{4+2}} = 1.77$$

∴ The polar plot is as under.

Example 7:

Sketch the polar-plot for

$$G(s) = \frac{10}{s^2\,(s+1)\,(s+2)}$$

Solution:

$$G(j\omega) = \frac{10}{(j\omega)^2\,(1+j\omega)(2+j\omega)}$$

$$\therefore |G(j\omega)| = \frac{10}{\omega^2\sqrt{1+\omega^2}\,\sqrt{4+\omega^2}}$$

and $\angle G(j\omega) = -180^\circ - \tan^{-1}\omega - \tan^{-1}\dfrac{\omega}{2}$

ω	$\lvert G(j\omega)\rvert$	$\angle G(j\omega)$	
0	∞	– 180°	→starting pt.
∞	0	– 360°=0°	→ending pt.
$\sqrt{2}$	1.18	– 270°	

The plot will intersect – 270°

i.e. 90° axis.

Hence intersection point can be obtained by using step [4.a]

$\angle G(j\omega) = -270^\circ$

$$= -180^\circ - \tan^{-1}\omega - \tan^{-1}(0.5\omega)$$

$$\therefore \tan^{-1}\omega + \tan^{-1}(0.5\,\omega) = 90^\circ$$

Solution gives $\omega = \sqrt{2}$

$$\therefore |G(j\omega)|_{\omega=\sqrt{2}} = \frac{10}{(\sqrt{2})^2\sqrt{1+2}\sqrt{(4+2)}} = 1.18.$$

Example 8:

(a) Sketch the polar-plot for $G(s) = \dfrac{k(1+T_1 s)}{(1+T_2 s)}$; $T_1 > T_2$.

(b) Repeat this problem for $T_1 < T_2$.

Solution:

$$G(j\omega) = \frac{k[1+j\omega T_1]}{[1+j\omega T_2]}$$

$$\therefore |G(j\omega)| = \frac{k\sqrt{1+\omega^2 T_1^2}}{\sqrt{1+\omega^2 T_2^2}}$$

and $\angle G(j\omega) = \tan^{-1}\omega T_1 - \tan^{-1}\omega T_2$

ω	$\lvert G(j\omega)\rvert$	$\angle G(j\omega)$	
	k	0°	→ Starting pt
* ∞	$k.\dfrac{T_1}{T_2}$	0°	→ End point
**1	$\dfrac{k\sqrt{1+T_1^2}}{\sqrt{1+T_2^2}}$	$\{\tan^{-1}T_1 - \tan^{-1}T_2\}$	

$$* \lim_{\omega\to\infty}|G(j\omega).| = \lim_{\omega\to\infty}\left\{\frac{k.\sqrt{1+\omega^2.T_1^2}}{1+\omega^2 T_2^2}\right\}$$

$$= \lim_{\omega\to\infty}\left\{\frac{k.\omega\sqrt{\dfrac{1}{\omega^2}+T_1^2}}{\omega\sqrt{\dfrac{1}{\omega^2}+T_2^2}}\right\} = k.\frac{T_1}{T_2}$$

As $T_1 > T_2$; $|G(j\omega)|_{\omega=\infty} > |G(j\omega)|_{\omega=0}$.

** $|G(j\omega)|$ for ω in between 0 and ∞ will also be in between

$|G(j\omega)|_{\omega=\infty}$ and $|G(j\omega)|_{\omega=0}$.

Also the angle will remain positive as $T_1 > T_2$.

Therefore the plot will be as under :

If $T_1 < T_2$ then after repeating above procedure the plot obtained.

***Example 9*:**

Sketch the polar-plot of $G(s) = \dfrac{\omega_n^2}{s^2 + 2\zeta\omega_n s + \omega_n^2}$

What will be the effect of ζ.

Solution:

$$G(j\omega) = \frac{\omega_n^2}{(j\omega)^2 + 2\zeta\omega_n \cdot j\omega + \omega_n^2}$$

$$= \frac{1}{1 - \left(\dfrac{\omega}{\omega_n}\right)^2 + j\dfrac{2\zeta\omega}{\omega_n}}$$

$$\therefore \quad |G(j\omega)| = \frac{1}{\left[\left\{1 - \left(\dfrac{\omega}{\omega_n}\right)^2\right\}^2 + \dfrac{4\zeta^2\omega^2}{\omega_n^2}\right]^{1/2}}$$

and $\angle G(j\omega) = -\tan^{-1}\left\{\dfrac{2\zeta\omega/\omega_n}{1-(\omega/\omega_n)^2}\right\} -$

$$\tan^{-1}\left\{\frac{2\zeta\omega}{\omega_n\sqrt{1-\left(\dfrac{\omega}{\omega_n}\right)^2}}\right\}$$

ω	$\|G(j\omega)\|$	$\angle G(j\omega)$	
0	1	1	→ Starting pt.
∞	0	– 180°.	→ Ending pt.
ω_n	$\dfrac{1}{2\zeta}$	– 90°	Crossing with –

90° axis.

Interaction with – 90° axis;

$$\angle G(j\omega) = -90^\circ = -\tan^{-1}\frac{2\zeta\omega}{\omega_n\sqrt{1-\left(\frac{\omega}{\omega_n}\right)^2}}.$$

$$\therefore \tan(-90^\circ) = \infty = \frac{2\zeta\omega}{\omega_n\sqrt{1-\left(\frac{\omega}{\omega_n}\right)^2}}$$

$$\therefore \omega_n\sqrt{1-\left(\frac{\omega}{\omega_n}\right)^2} = 0$$

As $\omega_n \neq 0$; $1-\left(\frac{\omega}{\omega_n}\right)^2 = 0$

$\therefore \omega = \omega_n$.

$$\therefore |G(j\omega)|_{\omega = \omega n} = \frac{1}{2\zeta}$$

The polar plot is shown in figure below.

Effect of ζ : As the value of ζ goes on increasing the intersection point $\left[\frac{1}{2\zeta}\right]$ will move upwards i.e. towards origin.

Example 10:

Sketch the polar-plot of transportion-lag.

Solution:

Transportation lag has T.F.;

$G(s) = e^{-Ts}$; where T = Time constant

$\therefore G(j\omega) = e^{-j\omega T} = \cos\omega T - j\sin\omega T$

$$\therefore |G(j\omega)| = \sqrt{\cos^2\omega T + \sin^2\omega T} = 1$$

$= \text{Constant}$

$$\angle G(j\omega) = \tan^{-1}\left\{\frac{-\sin\omega T}{\cos\omega T}\right\} = \tan^{-1}(-\tan\omega T)$$

$= -\omega T$ degrees

$\therefore \angle G(j\omega)$ will vary between 0 and 360°.

$\therefore$ Polar-plot will be a circle of unit radius as shown in figure below.

Example 11:

Sketch the polar-plot of

$$G(j\omega) = \frac{e^{-j\omega T_1}}{1 + j\omega T_2}$$

Solution:

$$|G(j\omega)| = \frac{1}{\sqrt{1+\omega^2 T_2^2}}$$

As $\left|e^{-j\omega T_1}\right| = 1$

$\angle G(j\omega) = -\omega T_1 - \tan^{-1} \omega T_2$

ω	$\|G(j\omega)\|$	$\angle G(j\omega)$	
0	1	0°	→ Starting pt.
1	$\frac{1}{\sqrt{1+T_2^2}}$	$-T_1 - \tan^{-1} T_2$	
10	$\frac{1}{\sqrt{1+100T_2^2}}$	$-10T_1 - \tan^{-1} 10T_2$	

→ Indicates that $\angle G(j\omega)$ will change between 0° and 360°.

→ Indicates that $|G(j\omega)|$ is reducing continuously; and finally as $\omega \to \infty$; $|G(j\omega)| \to 0$.

$\therefore$ The polar plot will be in the form of an spiral as shown in figure.

Example 12:

The Open Loop T.F. of a unity feed back system is

$$G(j\omega) = \frac{k}{j\omega(1 + j0.2\omega)(1 + j0.05\omega)}$$

(i) Compute ω_g, ω_p, GM and Φ_{PM} for k = 1 and comment on stability of the closed-loop system.

(ii) Find the maximum value of k to keep the system stable.

Solution:

Graphical Method :

(i) Sketch the polar-plot.

$$|G(j\omega)|_{k=1} = \frac{1}{\omega\sqrt{1+(0.2\omega)^2}\sqrt{1+(0.05\omega)^2}}$$

$\angle G(j\omega) = -90 - \tan^{-1}(0.2\omega) - \tan^{-1} 0.05\omega.$

ω	$\lvert G(j\omega)\rvert$ for k = 1	$\angle G(j\omega)$	
0	∞	-90°	→ Starting pt.
∞	0	-270°	→ End point
10	0.04	-180°	→ Intersec. with-180° axis.
1	0.97	-104°	

As the angle is changing from – 90° to – 270°; the plot will cross – 180°.

When polar-plot crosses – 180°; $\omega = \omega_p$.

$\therefore \angle G(j\omega)|_{\omega=\omega p} = -180^\circ.$

$= -90^\circ - \tan^{-1}(0.2\omega) - \tan^{-1}(0.05\,\omega)$

$\therefore \omega_p = 10$ rad/sec $\quad |G(j\omega)|_{\omega=\omega p,\ k=1} =$

$$\frac{1}{10\sqrt{1+(0.2\times10)^2}\sqrt{1+(0.05\times10)^2}} = 0.04$$

Let us take one more point say at $\omega = 1$

$\therefore |G(j\omega)|_{k=1} = 0.97;\ \angle G(j\omega) = -104^\circ$

The polar-plot is as under : [1 unit = 5cm]

a = 0.04

$$\therefore \text{GM} = \frac{1}{a} = \frac{1}{0.04} = 25$$

i.e. $\text{GM} = 20 \log\left(\frac{1}{a}\right) = 28$ db

Now $\omega_g = 1$ rad/sec

For finding the Φ_{PM} ; draw an arc with centre at 0 and radius = 1 unit. Let it cut the polar-plot at b.

$\therefore \Phi_{PM} = \theta^\circ = 76^\circ.$

(ii) Now we know that

k_{max} = Maximum gain to keep system in stable condition = $\frac{1}{a}$

$\therefore\ k_{max} = \frac{1}{0.04} = 25$

Analytical method

(i) Determination of ω_p :

At $\omega = \omega_p$; $\angle G(j\omega) = -180^\circ$

$\therefore -180^\circ = -90 - \tan^{-1}(0.2\omega) - \tan^{-1}(0.05\,\omega)$

$\therefore \omega_p = 10$ rad/sec

Determination of GM :

$$GM = \frac{1}{|G(j\omega)|_{\omega=\omega_p}}$$

$$\therefore GM = \frac{1}{\left[\frac{1}{10\sqrt{1+(0.2\times 10)^2}\,.\sqrt{1+(0.05\times 10)^2}}\right]} = \frac{1}{0.04} = 25$$

i.e. $GM = 20\log\left\{\frac{1}{|G(j\omega)|_{\omega=\omega_p}}\right\} = 28$db

$[GM]_{sp} = 40$ db

$\therefore\quad 40 = 20\log[GM]_{sp}$.

$\therefore [GM]_{sp} = 100$ Let k_2 be the gain

$$\therefore\ k_2 \times |G(j\omega)|_{\substack{\omega=\omega_p\\k=1}} = k_2 \times 0.04 = \frac{1}{100}$$

$\therefore k_2 = 0.25$

$\Phi_{sp} = 40^\circ$

Let k_3 be the gain and ω_{g3} be the new gain-cross over frequency.

We know that $\angle G(j\omega)|_{\omega g3} = \Phi_{sp} - 180^\circ$.

$\therefore -90 - \tan^{-1}(0.2\,\omega_{g3}) - \tan^{-1}(0.05\,\omega_{g3}) = 40 - 180^\circ = -140^\circ$

$\therefore \tan^{-1}(0.2\,\omega_{g3}) + \tan^{-1}(0.05\,\omega_{g3}) = 50^\circ$

Take tan on both sides

$\therefore \tan\{\tan^{-1} 0.2\,\omega_{g3} + \tan^{-1} 0.05\, w_{g3}\} = \tan 50^\circ$.

$$\therefore \frac{0.2\,\omega_{g3} + 0.05\,\omega_{g3}}{1-(0.2\times 0.05)\omega_{g3}} = 1.19$$

$$\therefore 0.25\,\omega_{g3} = 1.19 - 0.01\,\omega_{g3}^2$$

Solving this eq. and considering only positive value;

ω_{g3} = 4 rad/sec

We also know that

$$k_3 \times |G(j\omega)|_{\substack{\omega=\omega_{g3}\\ k=1}} = 1$$

$$\therefore k_3 = \frac{1}{10\sqrt{1+(0.2\times 4)^2}\,\sqrt{1+(0.05\times 4)^2}}$$

$\therefore k_3 = 5.2$

Determination of ω_g

At $\omega = \omega_g$; $|G(j\omega)| = 1$

$$\therefore \frac{1}{\omega_g\sqrt{1+(0.2\times\omega_g)^2}\,\sqrt{1+(0.05\times\omega_g)^2}} = 1$$

Solving by trial and error, we get :

$\therefore \omega_g$ = 1 rad/sec

Determination of Φ_{PM} :

$\Phi_{PM} = 180^\circ + \angle G(j\omega)\,|_{\omega=\omega g}$.

$\therefore \Phi_{PM} = 180^\circ - 90 - \tan^{-1}(0.2\times 1) - \tan^{-1}(0.05\times 1)$

$\therefore \Phi_{PM} = 180^\circ - 104^\circ$

$\therefore \Phi_{PM} = 76^\circ$

Stability : As GM and Φ_{PM} are positive; system is stable.

(ii) As per section 5.10;

$$k_{max} = \frac{1}{|G(j\omega)|_{\substack{\omega=\omega_p\\ k=1}}} = \frac{1}{0.04}$$

$\therefore k_{max} = 25$

Example 13:

Find Gain of the system to have GM = 10 db.

Solution:

We have obtained GM = 28 db i.e. 25 and $\Phi_{PM} = 76^\circ$

for k = 1 Also $|G(j\omega)|_{\substack{\omega = \omega p \\ k = 1}} = 0.04$

and $\angle G(j\omega)|_{\omega = \omega g} = -104^\circ$.

Let k_1 be the gain of the system to yield GM = 10 db

i.e. 10 db = 20 log GM

$\therefore$ GM = 3.16 units

(i) $k_1 \times |G(j\omega)|_{\substack{\omega = \omega p \\ k = 1}} = \dfrac{1}{[GM]_{sp}}$

$\therefore\ k_1 \times 0.04 = \dfrac{1}{3.16}$

$\therefore\ k_1 = 7.91$

Example 14:

The Open Loop T.F. of a unity feedback system is,

$$G(s) = \frac{k}{(1+s)\,(1+2\,s)(1+3\,s)}$$

(i) Find GM for k = 5

(ii) Find the value of gain to yield a GM of 20 db.

(iii) Find the maximum value of gain for stability of the system.

Solution:

(i) $|G(j\omega)| = \dfrac{k}{\sqrt{1+\omega^2}\,\sqrt{1+4\omega^2}\,\sqrt{1+9\omega^2}}$

and $\angle G(j\omega) = -\tan^{-1}\omega - \tan^{-1} 2\omega - \tan^{-1} 3\omega$.

At $\omega = \omega_p$; $\angle G(j\omega) = -180^\circ$

$\therefore\ -180^\circ = -\tan^{-1}\omega_p - \tan^{-1} 2\omega_p - \tan^{-1} 3\omega_p$

$\therefore\ \tan^{-1} 2\omega_p + \tan^{-1} 3\omega_p = 180 - \tan^{-1}\omega_p$.

Take tan

$$\therefore\ \frac{2\omega_p + 3\omega_p}{1 - 2\omega_p \times 3\omega_p} = \frac{0 - \omega_p}{1 + 0 \times \omega_p} = \frac{-\omega_p}{1}$$

$\therefore\ \dfrac{5\omega_p}{1 - 6\omega_p^2} = -\omega_p$ i.e. $1 - 6\omega_p^2 = -5$

$\therefore \omega = \pm 1$ Considering only +ve value.

$\therefore \omega_p = 1$ rad/sec

$$\therefore |G(j\omega)|_{\omega p} = \frac{k = 5}{\sqrt{1+1}\sqrt{1+4}\sqrt{1+9}} = \frac{5}{10}$$

$$\therefore |G(j\omega)|_{\omega p} = 0.5$$

$$\therefore \quad GM = \frac{1}{|G(j\omega)|_{\omega p}} = \frac{1}{0.5} = 2$$

i.e. GM = 6 db

(ii) $[GM]_{sp} = 20$ db

i.e. $20 = 20 \log [GM]_{sp} = 20 \log [GM]_{sp}$.

$\therefore [GM]_{sp} = 10$ units

Let k' be the gain.

We know that $k' \times |G(j\omega)|_{\substack{\omega = \omega p \\ k=1}} = \frac{1}{[GM]_{sp}}$

Now $|G(j\omega)|_{\substack{\omega = \omega p \\ k=1}} = \frac{1}{5}|G(j\omega)|_{\substack{\omega = \omega p \\ k=1}} = 0.1$

$\therefore k' \times 0.1 = \frac{1}{10} \quad \therefore \quad k' = 1$

(iii) $k_{max} = \frac{1}{|G(j\omega)|_{\substack{\omega = \omega p \\ k=1}}} = \frac{1}{0.1} = 10$

$\therefore \quad k_{max} = 10$

Example 15:

The Open Loop T.F. of a unity feedback system is;

$$G(s) = \frac{k(1+s)}{s^2[1+0.316s]}$$

(i) What is the maximum Φ_{PM} possible and at what frequency it will occur.

(ii) What should be the gain of the system for this Φ_{PM}.

Solution:

(i) $|G(j\omega)| = \frac{k\sqrt{1+\omega^2}}{\omega^2\sqrt{1+(0.316\omega^2)}}$

and $\angle G(j\omega) = -180 + \tan^{-1}\omega - \tan^{-1}(0.316\,\omega)$

$$\Phi_{PM} = 180 + \angle G(j\omega)\,|_{\omega=\omega p}$$

$\therefore \Phi_{PM} = \tan^{-1}\omega_p - \tan^{-1}(0.316\,\omega_p)$

Taking tan;

$$\therefore \tan\Phi_{PM} = \frac{\omega_p - 0.316\,\omega_p}{1+0.316\,\omega_p^2} = \theta\text{; say}$$

$$\theta = \frac{0.684\,\omega_p}{1+0.316\,\omega_p^2}$$

For maximum Φ_{PM}; $\dfrac{d\theta}{d\omega_p} = 0$

$\therefore 0.684\,(1 + 0.316\,\omega^2_p) - 0.684\,\omega_p \times 0.316 \times 2\,\omega_p = 0$

$\therefore 0.684 = 0.216\,\omega^2_p.$

$\therefore \quad \omega_p = 1.78$

$$\therefore [\Phi_{PM}]_{max} = \tan^{-1}\left\{\frac{\omega_p - 0.316\,\omega_p}{1+0.316\,\omega_p^2}\right\}$$

$$\therefore \quad [\Phi_{PM}]_{max} = 31.3^\circ.$$

(ii) At $\omega = \omega_p$; $|G(j\omega)| = 1$ $\quad \therefore \dfrac{k\sqrt{1+(1.78)^2}}{(1.78)^2\sqrt{1+(0.316\times 1.78)^2}} = 1$

$\therefore k = 1.781$

Example 16:

The Open Loop T.F. of a unity feedback system is

$$G(s) = \frac{5}{(1+s)\,(1+2s)}$$

Check the stability of the closed-loop system by Nyquist stability criterion.

Solution:

Let us construct the Nyquist plot [same as polar plot] of $G(j\omega)$:

$$|G(j\omega)| = \frac{5}{\sqrt{1+\omega^2}\,\sqrt{1+4\omega^2}}$$

& $\angle G(j\omega) = -\tan^{-1}\omega - \tan^{-1}2\,\omega.$

ω	$\lvert G(j\omega)\rvert$	$\angle G(j\omega)$	
0	5	0°	→ starting pt.
∞	0	-180°	→ end pt.
$\frac{1}{\sqrt{2}}$	1.85	-90°	→ crossing with imaginary axis

Similarly for $\omega = 0$ to $-\infty$ the polar-plot will be the mirror-image of polar plot for 0 to ∞. Hence for $-\infty < \omega < \infty$, the Nyquist plot will be as shown in figure below.

Stability checking : No. of poles on R.H.S. of $j\omega$ – axis of open loop T.F.=P=0.

Encirclement by Nyquist plot of $(-1 + j\,0)$ point in counter clockwise direction = N = 0.

As N = P; closed-loop system *will be stable.*

Example 17:

Check the stability of the system whose open-loop T.F. is,

$$G(s).H(s) = \frac{1}{s(s-1)}$$

Solution:

$$G(j\omega).H(j\omega) = \frac{1}{j\omega\,(j\omega-1)}$$

As there is one pole at origin, the Nyquist contour, T_{GH} of the given T.F. in S-plane will be as under :

I → Complete $j\omega$ – axis.

$\therefore\ s = j\omega;\ 0 < \omega < \infty.$

II → $s = R_e^{j\theta}\ R \to \infty$ and θ varies from 90° to -90° through 0°.

III → Complete – $j\omega$ axis. $\therefore\ s = -j\omega;\ \infty < \omega < 0$

IV → $s = \varepsilon\ .\ e^{j\theta}$.

$\varepsilon \to = 0$ and θ varies from $-9\ 0^o$ to $+9\ 0^o$ through 0°.

Now let us map this contour in the complex-plane.

Part I : $|G(j\omega).H(j\omega)| = \dfrac{1}{\omega\sqrt{1+\omega^2}}$ &

$\angle G(j\omega)\ .\ H(j\omega) = -90 + \tan^{-1}\omega.$

$\therefore$ at $\omega = 0$; $|G(j\omega) \, . \, H(j\omega)| = \infty$ and

$\angle G(j\omega) \, . \, H(j\omega) = -90^\circ + 180^\circ = 90^\circ$

at $\omega = \infty$; $|G(j\omega) \, . \, H(j\omega)| = 0$ and $\angle G(j\omega) \, . \, H(j\omega) = 180$

Part II : $s = R \, . \, e^{j\theta}$; $R \to \infty$;

θ varies from 90° to -90° through 0°.

$$\therefore \; G(s).H(s)\Big|_{s = Re^{j\theta}} = \lim_{R\to\infty} \frac{1}{R.e^{j\theta}[R.e^{j\theta} - 1]} = 0\,e^{-j2\theta}$$

for $\theta = 90^\circ$; $G(s) \, . \, H(s) = 0 \angle -180^\circ$.

for $\theta = -90^\circ$; $G(s) \, . \, H(s) = 0 \angle 180^\circ$.

$\therefore$ Its mapping will be a point at origin.

Part III : This will be the mirror-image of Part-I.

Part IV : $s = \varepsilon \, e^{j\theta}$;

$\varepsilon \to 0$ and θ varies from -90° to $+90^\circ$ through 0°.

$$\therefore \; G(s).H(s)\Big|_{s = \varepsilon e^{j\theta}} = \lim_{\varepsilon\to 0} \frac{1}{\varepsilon.e^{j\theta}[\varepsilon.e^{j\theta} - 1]} = \infty\,e^{-j\theta}$$

$\therefore$ When $\theta = -90^\circ$;

$G(s) \, . \, H(s) = \infty \angle +90^\circ$ and for $\theta = +90^\circ$

it will be $\infty \angle -90^\circ$

$\therefore$ Its mapping will be a circle of ∞ radius starting from -90° to $+90^\circ$.

$\therefore$ The complete Nyquist-plot in complex plane will be as under:

Check for stability :

P = No. of poles on RHS of $j\omega$ – axis of OPen Loop T.F. = 1

N = No of encirclement in counter-clock wise direction = 0.

As $N \neq P$ system will be unstable.

Example 18:

Check the stability of the system :

$$G(s) = \frac{k}{s(s+1)(s+2)}; \; H(s) = 1. \; k = 1$$

Solution:

The Open Loop T.F. has one pole at origin. Hence Nyquist contour in s-plane will be as shown. It has been divided into four parts as usual. Now let us discuss the mapping of each part :

Part I : $s = j\omega$

$$\therefore \; G(j\omega) = \frac{1}{j\omega\,(1+j\omega)\,(2+j\omega)}$$

ω	$\lvert G(j\omega)\rvert$	$\angle G(j\omega)$
0	∞	–90°
∞	0	–270°
$\sqrt{2}$	0.16	–180°.

∴ This will be the polar-plot of G(jω)

Part II : For this part $s = R_e^{j\theta}$; $R \to \infty$ and θ changes from 90° to –90° in clock-wise direction.

$$\therefore \; G(s)\Big|_{s=R.e^{j\theta}} = \lim_{R\to\infty} \frac{1}{Re^{j\theta}\,(Re^{j\theta}+1)(Re^{j\theta}+2)} = 0.e^{-j3\theta}$$

∴ at θ = 90°; G(s) = 0 ∠ – 270°

and at θ = – 90°; G(s) = 0 ∠ + 270°.

∴ This part will be mapped as a point at origin.

Part III : This will be the mirror-image of Part-I [as ω varies from 0 to – ∞].

Part IV : For this part; $s = \varepsilon\, e^{j\theta}$; $\varepsilon \to 0$;

θ changes from – 90° to + 90° in counter-clockwise direction.

$$\therefore \; G(s)\Big|_{s=\varepsilon e^{j\theta}} = \lim_{\varepsilon\to 0} \frac{1}{\varepsilon e^{j\theta}\,(\varepsilon e^{j\theta}+1)(\varepsilon e^{j\theta}+2)}$$

$$= \infty\, e^{-j\theta}.$$

∴ at θ = – 90°; G(s) = ∞ ∠ 90°

and at θ = + 90°; G(s) = ∞ ∠ – 90°

∴ This part will be mapped as a semi-circle of ∞ radius from + 90° to – 90° through 0°.

∴ This complete Nyquist stability plot will be as shown below:

Stability checking :

The no of poles on RHS of s-plane of G(s) = P = 0. The no of encirclement of (– 1 + j0) point in counter clockwise direction by Nyquist plot = N = 0.

As N = P;

∴ the system **will be stable.**

Example 19:

Repeat previous example given above for k = 10.

i.e. $G(s) = \dfrac{k}{s(s+1)(s+2)}$; H(s)=1; if k=10

Solution:

If k = 20; the crossing of Nyquist-plot with 180° – axis i.e. with –ve real axis will change. It will cross it at – 1.6.

Therefore – 1 + j 0 point will be encircled once in clockwise direction.

$\therefore$ N = – 1 $\therefore$ N ≠ P

$\therefore$ system will be unstable.

Example 20:

Check the stability of a unity feedback system Open Loop T.F. as

$G(s) = \dfrac{k}{s^2(s+4)}$; k = 1

Solution:

The Nyquist contour in s-plane will be as shown in below figure. Now let us map this in the complex plane.

Part I : s = j ω.

$$\therefore \; |G(j\omega)| = \frac{1}{\omega^2\sqrt{4+\omega^2}}$$

and $\angle G(j\omega) = -180^\circ - \tan^{-1}\dfrac{\omega}{4}$

ω	|G(jω)|	∠ G(jω)
0	∞	–180°
∞	0	–270°

Part II : s = $R_e^{j\theta}$; R → ∞ ; θ changes from 90° to – 90° in clockwise direction.

$$\therefore \; G(s)\Big|_{s=Re^{j\theta}} = \lim_{R\to\infty}\frac{1}{R^2 e^{j2\theta}(Re^{j\theta}+4)} = 0 \,.\, e^{-j3\theta}.$$

$\therefore$ It will be mapped as a point at origin.

Part III : s = – jω. This will be the mirror-image of part I.

Part IV : $s = \varepsilon e^{j\theta}$; $\varepsilon \to \infty$; θ changes from -90° to $+90^\circ$ in anti clockwise direction through 0°.

$$\therefore \quad G(s)\Big|_{s=\varepsilon . e^{j\theta}} = \lim_{\varepsilon \to 0} \frac{1}{\varepsilon^2 e^{j2\theta} (\varepsilon e^{j\theta} + 4)} = \infty\, e^{-j2\theta} .$$

$\therefore$ at $\theta = -90^\circ$ and $G(s) = \infty \angle 180^\circ$ and

at $\theta = +90^\circ$; $G(s) = \infty \angle -180^\circ$

$\therefore$ This part will be mapped as a circle of ∞ radius from 180° to -180° through 0°. Therefore the complete Nyquist stability plot will be as shown below the figure.

Stability checking : $P = 0$; $N = -1$

$N \neq P$ $\therefore$ System is unstable.

Example 21(a):

The Open Loop T.F. of a unity feedback system is;

$$G(s) = \frac{(4s+1)}{s^2 (s+1)(2s+1)}$$

Check the stability of the system.

Solution:

The Nyquist contour in s-plane is as shown in figure. The various parts will be mapped into complex-plane as under :

Part I : $s = j\omega$.

$$\therefore \quad G(s) = \frac{1+4j\omega}{(j\omega)^2 (1+j\omega)(1+2j\omega)}$$

$$\therefore \quad |G(j\omega)| = \frac{\sqrt{1+16\omega^2}}{\omega^2 \sqrt{1+\omega^2}\sqrt{1+4\omega^2}}$$

and $\angle G(j\omega) = -180^\circ - \tan^{-1}\omega - \tan^{-1} 2\omega + \tan^{-1} 4\omega$.

ω	$\lvert G(j\omega)\rvert$	$\angle G(j\omega)$	
0	∞	-180°	$\to$ starting pt.
∞	0	-270°	$\to$ end pt.
0.354	10.6	180°	$\to$ crossing with –ve real - axis.

* When $G(j\omega)$ – plot intersects –ve real axis; $\angle G(j\omega) = 180^\circ$

$\therefore\ 180^\circ = -180^\circ - \tan^{-1}\omega - \tan^{-1}2\omega + \tan^{-1}4\omega.$

$\therefore\ \omega = 0.354$ rad/sec.

Part II : As usual it will be mapped as a point at origin.

Part III : It will be the mirror-image of part-I.

Part IV : $s = \varepsilon\, e^{j\theta}$; $\varepsilon \to 0$ &

θ changes from -90° to $+90^\circ$ through 0°.

$$\therefore\ G(s)\Big|_{s=\varepsilon e^{j\theta}} = \lim_{\varepsilon\to 0}\frac{(4\varepsilon e^{j\theta}+1)}{\varepsilon e^{j2\theta}(\varepsilon e^{j\theta}+1)(2\varepsilon e^{j\theta}+1)} = \infty . e^{-j2\theta}$$

$\therefore$ at $\theta = -90^\circ$; $G(s) = \infty \angle +180^\circ$

and at $\theta = 90^\circ$; $G(s) = \infty \angle -180^\circ$

$\therefore$ This part will be mapped as a circle of ∞ radius from $+180^\circ$ to -180° through 0°.

The complete Nyquist stability plot will be as upper.

Stability checking : $P = 0$; $N = -2$,

$\because\ P \neq N$;

$\therefore$ System is unstable.

Example 21(b):

Sketch the asymptotic Bode plot for

$$G(s) = \frac{10s}{(s+5)}$$

Solution:

Bring G(s) to standard form

$$G(s) = \frac{10s}{(s+5)} = \frac{10s}{5\left(1+\frac{s}{5}\right)} = \frac{2s}{\left(1+\frac{s}{5}\right)}$$

$$G(j\omega) = \frac{2(j\omega)}{\left(1+\frac{j\omega}{5}\right)}$$ Where $s = j\omega$.

Step 2 : Write down each factor in a table and write remarks. The following factors are present.

(1) Constant factor : 2

(2) Zero at origin of order 1 : $j\omega$.

(3) Finite pole factor : $\dfrac{1}{1+\dfrac{j\omega}{5}}$

The magnitude and phase remarks are as follows :

No.	Factor	Magnitude Curve	Phase Curve
1.	2	Horizontal line of magnitude 20 $\log_{10}$ |2| = 6.02 db for all ω.	Horizontal line at θ = 0° for all ω.
2.	jω	Straight line of slope 20 db/dec, having magnitude 0 db at ω = 1	Horizontal line at θ = 90° for all ω.
3.	$\dfrac{1}{1+\dfrac{j\omega}{5}}$	(1) 0 db line for ω < 5 (2) Line at slope $-\dfrac{20\,\text{db}}{\text{dec}}$ for all ω > 5	(1) Line at 0° for ω < $\dfrac{5}{5}$ i.e.ω < 1 (2) Line at – 9 0° for ω > 5 × 5 i.e. ω > 25

Step 3 : Draw the magnitude plot for each factor clearly.

All these are drawn in figure on next page.

Note :

(1) For factor k = 2, a line at 6 db is drawn horizontally.

(2) For jω a point at – 20 db at ω = 0.1 and 0 at ω = 1 are joined, to get a line at a slope of 20 db/dec. When extended at ω = 10 will touch 20 db indicating 20 db/dec slope.

(3) For ω < 5 slope is 0 db horizontal. For ω > 5 slope is – 20 db/dec. Take point ω = 50, draw a line from ω = 5, 0 db to ω = 50 and point – 20 db. So slope is – 20 db/dec. Extend it (Check : At ω = 500 the line will have magnitude – 40 db).

Step 4 : Add all points clearly.

(1) Proceed from left most point. At first we have a step of level 6 db and a ramp of slope + 20 db/dec. This is case 2 addition. Addition is a ramp of slope 20 db/dec and raised by a value equal to step value i.e. 6 db. Since at ω = 0 ramp value was 0 now after addition it will

be 6 db. Thus line of slope 20 db/dec with a value at $\omega = 1$ of 6 db is resultant of curve (1) and (2). Draw this line upto $\omega = 5$. Now curves (1) and (2) need not be taken individually but their resultant is to be counted only.

(2) At $\omega = 5$, there is a down slope line of slope – 20 db/dec.

Also the resultant of (1) and (2) is an upslope line of value 20 db/dec. [At $\omega = 5$ this resultant has a 21 db value (15 db + 6 db = 21 db)]. This corresponds to ramp addition of case 3. Since slopes are equal and opposite the resultant has 0 slope i.e. horizontal line at 21 db from $\omega > 5$ onwards. This is the resultant of [(1) + (2) + (3)]. All factors are considered. The resultant is shown dotted.

Step 5 : Draw the phase plot clearly for each factor.

Action :

(1) For factor k = 2, phase angle is at $\theta = 0°$. Draw it (X-axis itself).

(2) For factor $j\omega$, phase angle is $\theta = 90°$

(3) For factor $\dfrac{1}{1+j\dfrac{\omega}{5}}$ phase angle θ is $\theta = 0°;\ \omega < 1$

$= -90°;\ \omega > 25$

and linear between 0 and – 90° for $1 < \omega < 5$. All these are drawn.

Step 6 : Add all points clearly and plot resultant curve.

(1) For all $\omega < 1$ curves (1) and (3) give phase of 0° and curve (2) has a phase of 90°. So resultant is 90°. Only curve (2).

(2) For all $\omega > 25$ curve (1) gives 0°, curve (2) 90° and curve (3) – 90°.

So resultant is 0°.

(3) For $1 < \omega < 25$ there is a linear change from a value 90° to 0°. This is plotted to get the phase plot shown dotted.

Note :

(1) If the student can draw the magnitude addition without reference to case additions, as long as the result is same, any other method (if any) is valid.

(2) The exact smooth curve may be drawn if asked, what we have drawn are asymptotic approximations.

(3) In the Bode curves, the resultant is shown dotted at an angle (slightly) less than 90°) for $\omega < 1$; only to indicate clearly.

(The dotted resultant is same as the dark line itself which should be noted.)

(4) All slops are in db/dec.

Example 22:

Construct the asymptotic Bode plot for

$$G(s) = \frac{4\left(1+\frac{s}{2}\right)}{s^2\left(1+\frac{s}{8}\right)\left(1+\frac{s}{10}\right)}$$

Solution:

Step 1 : The function is already in standard form

Put s = jω.

$$G(j\omega) = \frac{4\left(1+j\frac{\omega}{2}\right)}{(j\omega)^2\left(1+\frac{j\omega}{8}\right)\left(1+j\frac{\omega}{10}\right)}$$

Step 2 : The following factors are present:

(1) Constant factor : 4

(2) Pole at origin of order 2 : $(1/j\omega)^2$

(3) Finite poles at : $\dfrac{1}{1+j\frac{\omega}{8}}$

(4) Finite Pole at : $\dfrac{1}{1+j\frac{\omega}{10}}$

(5) Finite Zero at : $\dfrac{1}{1+j\frac{\omega}{2}}$

Step 2 : Draw the magnitude plot (fig 13.83). Note that a slope of 20 db/dec or 40 db/dec are drawn as said in previous problem.

Step 3 : Upto ω = 2 only curves (1) and (2) are present the resultant of curve (1) and (2) is as per case 2 addition a ramp of slope 40 db/dec of value at ω = 1 is (0 + 10) = 12 db.

This curve comes upto ω = 2. At ω = 2 onwards a + 20 db/dec slope begins. So the resultant has a slope of 20 db/dec of value of (– 40 + 20 =

20 db/dec.) Here point of addition is not known. Ramp has slope 0 db at ω = 1. Shift by amount 12 db here to get resultant. Incidentally the line slopes at – 20 db/dec from ω = 2 to ω =8 and at ω = 8 the – 20 db/dec adds to this. This is case 3 addition.

And resultant slopes at – 40 db/dec from ω = 8. From ω = 8 to ω = 10, this continues. And at ω = 10, – 20 db/dec slope again adds up, make a slope at – 60 db/dec. (To draw lines quickly at – 20, – 40, –60 db/dec etc. draw in pencil between any decade points in rough 20 db, 40 db, etc lines and carefully shift the slope with a moving rule !).

Step 5 : Draw the phase plot.

Step 6 : From all – ∞ < ω < 0.4 the phase angle is – 180°. From 0.4 to 10 the angle would have reduced by 90° due to (5).

Draw a line such that it ends at – 90° from – 180° to 0.4. But at 1.6 due to (3),this decrease is prevented so draw the above line only upto ω = 1.6.

At ω = 1.6 a downward slope cancels the upward slope and maintains same level till ω = 2. This is by graph – 144°.

From ω – 2 to ω = 50 a linear slope changing the phase angle by adding – 90° to – 144° takes place. Draw this, from ω = 50. No more additions take place and no slope present. So the resultant remains at – 90° + (– 144°) = – 234°.

Example 23:

Draw the Bode plots for the following open loop transfer function. Indicate the gain and phase margins.

$$G(s) = \frac{1}{s(s+1)(s+5)}$$

Solution:

Step 1 : $$G(s) = \frac{1}{s(s+1)(s+5)} = \frac{1}{s\left(1+\frac{s}{1}\right)\left(1+\frac{s}{5}\right)5}$$

$$= \frac{0.2}{s\left(1+\frac{s}{1}\right)\left(1+\frac{s}{5}\right)} = \frac{0.2}{j\omega\left(1+\frac{j\omega}{5}\right)\left(1+\frac{j\omega}{5}\right)}$$

Step 2 : Draw the magnitude plot (See figure below.)

Step 3 : Upto $\omega = 1$ only curve (1) and (2) are present. Case 1 addition gives resultant at – 20 db upto $\omega = 1$ shifted by – 14 db at $\omega = 1$ (0 + (–14)) db at $\omega = 1 = -14$ db). At $\omega = 1$, the slope changes to –40 db resultant due to (3). At $\omega =$ slope changes to – 60 db, again by case 3 addition and continuous unchanged.

Step 4 : Draw the phase plot.

Step 5 : For all $\omega < 0.2$ $\Phi = -90^\circ$ [(1) + (2)]. From 0.2 to $\omega = 1$, Φ varies linearly and towards – 180° but for term (4). Join points from – 90° and – 180° at $\omega = 5$ but draw curve only till $\omega = 1$. At $\omega = 1$, $\Phi = -135^\circ$. From $\omega = 1$ to $\omega = 5$ the two downward slopes add up. At $\omega = 5$ the $\Phi = -225^\circ$. From $\omega = 5$ to $\omega = 25$ again single slope upto –270°. And from $\omega = 25$ slope is – 270°. This is phase plot.

On stability

Step 6 : Unity gain occurs at (from magnitude curve where resultant crosses 0 db at w = 0.2. This is gain cross over frequency.)

$\therefore \omega_{gc} = 0.2$

Step 7 : Corresponding phase at $\omega_{gc} = 0.2$ is – 90°.

$\therefore$ Phase margin = $180 + \Phi = 180^\circ + (-90^\circ) = 90^\circ$.

Step 8 : Phase plot has phase of – 180° at $\omega = 2$. This is phase crossover frequency $\omega_{pc} = 2$.

Step 9 : The value at $\omega_{pc} = 2$ of the magnitude is - 28 db approximately from graph i.e. 28 db below 0 db. Hence gain margin = 28 db. Since both gain and phase margin have positive values system is stable.

Example 24:

We have,

Solution:

Low frequency slope is + 40 db/dec. This is a case 2 type problem.

Step 1 and 2 : Not applicable.

Step 3 : As slope is +ve it is zero at origin order is

$$= 2\left[2 \times (20)\frac{\text{db}}{\text{dec}} = 40\right], \text{ the term is } (j\omega)^2$$

Step 4 : At $\omega = 1$ position value of k is to be calculated as follows.

At $\omega = 4$ position value is 0 db.

Slope is + 20 db/sec. $\omega = 10^3$ and $\omega = 10^4$ are a decade apart, db value at ω = 3 is 20 db. It remains at ω = 2 also. Now ω = 2 and 1 are 1 decade apart. Line slopes at 40 db/dec.

∴ At ω = 1, is 40 db below value at ω=2

∴ At ω = 1 20 – (40) = – 20 db.

∴ 20 log k = – 20

∴ k = 0.1

Step 5 to 8 : Change in slope occurs at log ω = 2, 3

∴ a = 10^2, 10^3, following table gives the results.

	ω = 2	ω = 3
Step 5 :	a = 10^2 = 100	a = 10^3 = 1000
Step 6 :	0 – (+ 40) db = –40 db/dec	– 20 db – 0 = – 20 db/dec
Step 7 :	n = 2	n = 1
Step 8 :	$\frac{1}{\left(1+j\frac{\omega}{100}\right)^2}$	$\frac{1}{\left[1+j\frac{\omega}{10^3}\right]}$

$$\therefore \ T(s) = \frac{0.1\,\omega^2}{\left(1+j\frac{\omega}{100}\right)^2\left(1+j\frac{\omega}{10^3}\right)}$$

Example 25:

Follow asymptotic approximation and determine the transfer function from the following test data. What is the type and order of the system :

Gain (db) : 34 28 24.6 14.2 8 1.5 –3.5 –7.2

Frequency (rad/sec) : 0.1 0.2 0.3 0.7 1.0 1.5 2.0 2.5

Gain –12.5 –14.7 –16 –17.5 –17.5 –17.5

Frequency : 4 5 6 9 20 25

Solution:

The results should be plotted on semilog paper.

Maximum db value = 34 and min – 17.25

Choosing a scale 1 cm = 10 db

Section	Slope	Change At
AB	– 20 db/dec	—
BC	– 40 db/dec	0.3
CD	– 20 db/dec	4
DE	0 db	9

The graph is not drawn and the student should draw it and then check further with us. From the graph it is clear there are four sections.

The value of resultant without correction at ω = 1 is 20 log k = 8

From which k = 2.5

Type of the system : One and Order of the system : Two

The transfer function is

$$= \frac{2.5\left(1+\frac{s}{4}\right)\left(1+\frac{s}{9}\right)}{s\left(1+\frac{s}{0.3}\right)} \quad \text{...(Ans.)}$$

Number of open-loop poles in right halfs-plane

Example 26:

For a second order system with unity feedback,

$$G(s) = \frac{200}{s(s+8)}$$

Find the frequency domain specifications

Solution:

The transfer function is

$$T(s) = \frac{C(s)}{R(s)} = \frac{G(s)}{1+G(s)H(s)};$$

H(s) = 1

$$T(s) = \frac{\frac{200}{s(s+8)}}{1+\frac{200}{s(s+8)}} = \frac{200}{s^2+8s+200}$$

Comparing with canonical transfer function $\frac{\omega_n^2}{s^2+2\zeta\omega_n s+\omega_n^2}$

$\omega_n^2 = 200$ $2\zeta\omega_n = 8$

$\therefore \zeta = 0.283$

$$M_r = \frac{1}{2\zeta\sqrt{1-\zeta^2}}$$

Substituting values of ζ

$M_r = 1.842$

$$\omega_r = \omega_n\sqrt{1-2\zeta^2}$$

Substituting the values

$\therefore \omega_r = 12.96$ rad/sec

$$\omega_b = \omega_n\left\{(1-2\zeta^2)\pm\sqrt{4\zeta^4-4\zeta^2+2}\right\}^{\frac{1}{2}}$$

Substituting for ω_n and ζ

$\omega_b = 14.14\ \{0.893+1.31\} = 14.14\times1.46$

$\therefore \omega_b = 20.73$ rad/sec.

Example 27:

Consider feedback system shown in the figure. Find k and a to satisfy $M_r = 1.04$ and $\omega_r = 11.55$ rad/sec. (b) For the values of k and a obtained in part (a), determine the settling time, peak over-shoot and band-width.

Solution:

(a) $M_r = \dfrac{1}{2\zeta\sqrt{1-\zeta^2}} = 1.04$

Squaring $1.018 = \dfrac{1}{4\zeta^2(1-\zeta^2)}$

Subtitute $x = \zeta^2$

$x^2 - x + 0.231 = 0 \Rightarrow x = 0.6375$

or $x = 0.3625$

$\therefore\ \zeta = \sqrt{x} = 0.7984$ or 0.6020

As M_r occurs for $\zeta \le \dfrac{1}{\sqrt{2}}$ only i.e. for $\zeta \le 0.707$;

$\zeta = 0.602$ will only be considered

$$\omega_r = \omega_n\sqrt{1-2\zeta^2} = 11.55$$

$\omega_n = 22.04$

Characteristic equation of the given system i.e.

$$1 + G(s)\,H(s) = 0 \text{ is}$$

$$1+\frac{k}{s(s+a)} \text{ i.e. } s^2 + as + k = 0$$

Comparing if with the standard form i.e.

$$s^2 + 2\,\zeta\,\omega_n + \omega^2_n = 0$$

$$k = \omega^2_n = 485$$

$$2\,\zeta\,\omega_n = a = 2 \times 0.602 \times 22.04 = 25.33$$

(b) $$t_s = \frac{4}{\zeta\omega_n} = \frac{4}{0.602\times 22.04} = 0.3 \text{ sec}$$

$$M_p = e^{-\pi\zeta/\sqrt{1-\zeta^2}} = e^{-\pi\times 0.602/\sqrt{1-(0.602)^2}}$$

$$= 0.0936 = 9.36\%$$

ω_b = Band-width

$$= \omega_n\left[(1-2\zeta^2)+\sqrt{(1-2\zeta^2)^2+1}\right]^{1/2}$$

= 25 rad/sec

Example 28(a):

The transient-response of a second-order under damped system subjected to a unit step input is found to have a peak overshoot of 16.2% at time $t = \frac{\pi}{5\sqrt{3}}$ sec.

If this system is subjected to a sinusoidal input; determine

(a) frequency of input at which the amplitude of steady-state response will be maximum.

(b) The maximum value of steady-state output.

Solution:

Given % $M_p = 16.2 = 100\, e^{\frac{-\pi\zeta}{\pi-\zeta^2}}$

$\therefore \zeta = 0.5$

Also given; $t_p = \frac{\pi}{\omega_d} = \frac{\pi}{5\sqrt{3}}$

$$\therefore \quad \omega_d = 5\sqrt{3} = \omega_n\sqrt{1-\zeta^2}$$

$$\therefore \quad \omega_n = 10$$

(a) ω_r = The frequency at which maximum output occurs

$$= \omega_n\sqrt{1-2\zeta^2} = 7.07 \text{ rad/sec}$$

(b) $$M_r = \frac{1}{2\zeta\sqrt{1-\zeta^2}} = 1.15$$

Example 28(b):

A sinusoidal input signal is applied to a C.S. The value of maximum magnitude obtained from such a system is twice the input amp. occuring at a frequency of 6 rad/sec. Calculate the damping ratio and its net frequency.

Solution:

$$M_r = 2 \text{ units} = \frac{1}{2\zeta\sqrt{1-\zeta^2}}$$

$\therefore\ \zeta = 0.258$ and 0.966 As $\zeta > 0.707$

$\zeta = 0.258$ should be considered.

$\therefore\ \zeta = 0.258$

$$\omega_r = \omega_n\sqrt{1-2\zeta^2} = 6$$

$\therefore\ \omega_n = 6.208$ rad/sec

Example 28(b):

A unit-step response data of a 2nd order system is given below. Obtain the corresponding frequency response indices (M_r ω_r, ω_b) for the system.

t(sec) 0, 0.05, 0.1, 0.15, 0.2, 0.25, 0.3, 0.35, 0.4, 0.45, 0.5

c(t) 0, 0.25, 0.8, 1.08, 1.12, 1.02, 0.98, 0.98, 1, 1, 1

Solution:

M_p - 0.12 $\therefore$ find ζ

From the given data;

$M_p = 0.12 = 12\%$ (occuring at t=0.2 sec)

and $t_s = 0.25$ sec

(as the response is setting within ± 2% of output).

$$\therefore M_p = 0.12 = e^{\frac{-\pi\zeta}{\sqrt{1-\zeta^2}}}$$

$$\therefore \zeta = 0.55 \quad t_s = \frac{4}{\zeta\omega_n}$$

$$= 0.25 \text{ rad/sec}$$

$$\therefore \quad \omega_n = 29.09 \text{ rad/sec}$$

Calculation of frequency response specifications

$$M_r = \frac{1}{2\zeta\sqrt{1-\zeta^2}} = 1.09$$

$$\omega_r = \omega_n\sqrt{1-2\zeta^2} = 18.28 \text{ rad/sec}$$

$$\omega_b = \omega_n\left\{(1-2\zeta^2)+\sqrt{2-4\zeta^2+4\zeta^4}\right\}^{\frac{1}{2}}$$

$$= 29.09\ \{0.395 + 1.156\}^{1/2}$$

$$= 36.23 \text{ rad/sec.}$$

Example 28(c):

Derive the correlation between phase-margin and damping factor.

Solution:

Consider a standard 2nd order system.

The open loop T.F. =

$$G(j\omega).H(j\omega) = \frac{\omega_n^2}{(j\omega)(j\omega+2\zeta\omega_n)}$$

At gain-cross-over frequency (ω_g) :

$$|G(j\omega)|\ .\ |H(j\omega)| = 1$$

$$\therefore \left|\frac{\omega_n^2}{j\omega_g(j\omega_g+2\zeta\omega_n)}\right| = 1$$

$$\text{i.e.} \quad \frac{\omega_n^2}{\omega_g\sqrt{\omega_g^2+4\zeta^2\omega_n^2}} = 1$$

$$\therefore \omega_g^4 + 4\zeta^2\omega_n^2.\omega_g^2 - \omega_n^4 = 0$$

If say $x = \omega^2_g$ is substituted; it becomes a quadratic eq.

$\therefore$ Solution given;

$$\omega_g^2 = \frac{-4\zeta^2\,\omega_n^2 \pm \sqrt{16\,\zeta^4\,\omega_n^4 + 4\,\omega_n^4}}{2}$$

$$= -2\zeta^2\,\omega_n^2 \pm \omega_n^2\sqrt{4\zeta^4+1}$$

$$\therefore \left(\frac{\omega_g}{\omega_n}\right)^2 = -2\zeta^2 + \sqrt{1+4\zeta^4}$$

Now

$$\Phi_{PM} = 180^\circ + \angle G_{(j\omega)}.H_{(j\omega)}\Big|_{\omega=\omega_g}$$

$$\Phi_{PM} = 180^\circ - 90^\circ - \tan^{-1}\left(\frac{1}{2\zeta}\cdot\frac{\omega_g}{\omega_n}\right)$$

$$\therefore \Phi_{PM} = 90^\circ - \tan^{-1}\frac{\left\{\{-2\zeta^2+\sqrt{1+4\zeta^4}\}^{\frac{1}{2}}\right\}}{2\zeta}$$

$$= 90 - \theta \qquad \text{...(say)}$$

When ζ is plotted for various values of ζ [$\zeta < 0.707$]; the graph is in under.

A reasonably good linear approximation is $\zeta \approx 0.01\,\Phi_{PM}$; Φ_{PM} is in degrees.

$$\therefore \tan\Phi_{PM} = \tan(90-\theta) = \frac{1}{\tan\theta}$$

$$\therefore \tan\Phi_{PM} = \frac{2\zeta}{\left\{-2\zeta+\sqrt{1+4\zeta^4}\right\}^{\frac{1}{2}}}$$

$$\therefore \Phi_{PM} = \tan^{-1}\left\{\frac{2\zeta}{\left(-2\zeta^2+\sqrt{1+4\zeta^4}\right)^{\frac{1}{2}}}\right\}$$

Example 29:

Sketch the polar-plot of following system:

$$G(s) = \frac{10}{s(s+1)}$$

Solution:

Given $G(j\omega) = \frac{10}{j\omega(1+j\omega)}$

$\therefore |G(j\omega)| = \frac{10}{\omega\sqrt{1+\omega^2}}$ and

$\angle G(j\omega) = -90 - \tan^{-1}\omega.$

The polar plot is shown in Figure

ω	$\lvert G(j\omega)\rvert$	$\angle G(j\omega)$
0	∞	-90° →starting pt.
∞	0	-180° →ending pt.
1	7	-135°
10	0.1	$\approx -180^\circ$.

4

Classification of Control System

NTRODUCTION

Broadly the control system are classified into two categories :

(i) Open - Loop control systems

(ii) Closed - Loop control systems.

[A] Open - Loop Control system : It is system which does not automatically correct it's input for variations in its output. This system is represented as under.

• **Controller :** It is the device used for controlling the quantity of interest in the process.

• r → Reference input signal

• u → Controlling signal

• c → Output

Examples :

(i) **Speed Control of a D.C. Shunt Motor :** This can be controlled either by armature voltage control or field current control. Let us say that field control method is used and the desired speed is 1400 rpm. A tachometer is being used for measuring the speed. If due to some disturbance speed drops to 1300 rpm, then the speed can not be brought back to 1400 rpm unless we manually adjust the speed by field rheostat. Hence the control is not automatic.

(ii) **Traffic Control System :** When the traffic control signals change only with time it is open loop system as they are not changing according to the flow of traffic. If from any particular side traffic flow ends in 10 sec & signal is set for 20 sec, then traffic from next side has to wait unnecessarily for next 10 sec.

[B] Closed Loop Control System : It is a system in which the output signal has a direct effect on the control action. This system is also called as *'Automatic Control System'* as control action is automatic. In this system the output is fed-back for comparison with the reference input to get appropriate control action, hence it is also called as *'Feedback Control System'*. It is represented as under.

- b → Feed - back signal
- e → Error signal

Examples :

(i) **Speed Control of D.C. Shunt Motor:** In above example if the tachometer is used not only as a measuring device but also as a feed back element, then the input to the controller will get adjusted automatically so that speed remains constant to 1400 rpm.

(ii) **Traffic Control System :** In above example if a sensor is used on each side for detecting the flow of traffic then as soon as traffic signal ends from one side, the traffic flow from the next side will be started.

Hence by using tachometer & traffic sensor as feedback elements the systems can be automatized.

Types of Feed Back Control System

They are classified in following ways :

(1) *According to the nature of variation of output :*

(i) **Linear C.S. :** When output follows the principle of superposition.

(ii) **Non-linear C.S. :** When output does not follow the principle of superposition.

In practice all system contain non-linear characteristics, but for analysis purpose many system are treated as linear system.

(2) *According to the nature of control signal:*

(i) **Continuous Time Systems :** In this system, the signal is the continuous function of time variable.

(ii) **Digital or Discrete [Sampled] Data C.S. :** In this system, the signal at one or more points of time is either in the form of a pulse train or a digital code.

(3) *According to the nature of parameter variation :*

(i) **Time Variant C.S. :** In this system, the parameters of the system vary with time.

(ii) **Time in-variant C.S. :** In this system, the parameters of the system do not vary with time.

(4) *According to input and output :*

(i) **Single Input Single Output [SISO] System :** In this system, only one variable is controlled by one input.

(ii) **Multi Input Multi Output [MIMO] System :** When more than one variable of the system are to be controlled with a number of inputs, system is called as MIMO System.

(5) *According to the nature of feed back signal :*

(i) **Positive Feedback [Re-genarative feedback] C.S. :** When feedback signal is additive [+]; it is positive feedback system.

(ii) **Negative Feedback [Degenerative feedback] C.S. :** When feedback signal is substractive [–]; it is negative feedback system.

(6) *According to the purpose of the system:*

(i) **Regulator :** If the object of the C.S. is to maintain some physical variable at some constant value in the presence of some disturbances, the system is called as Regulator; e.g. a voltage stabilizer has desired output which remain constant.

(ii) **Servo - Mechanism :** If the object of the C.S. is that the physical variable should follow or track some desired time function, the system is called as Servo mechanism. It is a system which controls the mechanical position or its time derivative; e.g. Servo-voltage stabilizer in which position of secondary tap is controlled using Servo-motor.

CONTROL SYSTEM

It is the means by which any quantity of interest in a process is maintained or altered in the desired manner. To decide the control system for any process we must know two things, viz

(i) Quantity of interest in the process

(ii) The manner in which it has to be maintained or altered.

MATHEMATICAL MODELLING

Any system has two different aspects; Analysis & Design. By analysing any system one can get the idea about the dynamic behavior of the system

Therefore the analysis of control system is essential. This is done by following steps.

Physical System

↓

Physical Model

↓

Mathematical Model

↓

Transfer Function;
Block-Diagram
Signal Flow Graph

↓

Mathematical Solution
⇒ Dynamic Behaviour

Explanation : 'Physical system' is a collection of physical objects connected together to serve an object. When idealized assumptions are made for analysis or synthesis it is called as *'Physical Model'*. It can be modelled in a number of ways depending on the problem to be solved & accuracy required. *'Mathematical Model'* gives a clear understanding of the system in terms of cause & effect relationship among the system components.

The convenient representation of the mathematical model is by 'Transfer Function' or 'Block-Diagram' or 'Signal Flow Graph' which gives the relationship between output & input in terms of system parameters. The mathematical tools or conceptual aspects are applied to this model in order to get 'Mathematical Solution' which gives the 'Dynamic Behavior' of the system. Finally this solution is translated to the 'Physical System'.

The mathematical model depends on the physical system; but the techniques of T.F., B.D. & S.F.G are same. Hence first we must get convenient with these techniques.

SYNCHROS

It is a rotary device which operates on transformer principle & produces a correlation between an angular position & a voltage or a set of voltages. The input is the angular displacement of the shaft & output is the voltage. It consists of two units viz. synchro transmitter & synchro control transformer.

Synchro - Transmitter : The construction is similar to that of a 3 - phase alternator. Three identical coils are placed in the stator slots and they are displaced by 120°. The rotor is dumb - bell shaped & a a.c. voltage is applied across its winding.

Observe the synchro-transmitter alone.

Let

$v_r = V_{rm} \sin \omega t$ = Voltage applied to the rotor winding. ...(1)

θ = Angular displacement of rotor with respect to the axis of s_2 winding.

Due to transformer action voltages will be induced in the stator windings; say v_{s1}, v_{s2} & v_{s3}.

These are phase voltage

$$\therefore \left.\begin{aligned} v_{s2} &= K\, v_r \,.\cos\theta \\ v_{s1} &= K\, v_r \cos(\theta + 120°) \\ v_{s3} &= K\, v_r \cos(\theta - 120°) \end{aligned}\right\} \quad ...(2)$$

$V_{s1\,s2}$ = Voltage between s_1 & s_2 terminal

$$= \overline{V}_{s1} - \overline{V}_{s2}$$

$$V_{s1\,s2} = \sqrt{3}\ K\ v_r\ [\cos(\theta + 120) - \cos\theta]$$

$$= \sqrt{3}\ K\ V_{rm} \sin(\theta - 120) \sin \omega t$$

Similarly,

$$V_{s2\,s3} = \sqrt{3}\ K\ V_{rm} \sin(\theta + 120) \sin \omega t$$

$$V_{s3\,s1} = \sqrt{3}\ K\ V_{rm} \sin\theta \,.\, \sin \omega t \quad ...(3)$$

When $\theta = \theta$ i.e. if rotor axis & s_2 axis are in the same line;

$v_{s2} = K\, v_r$ & $V_{s3\,s1} = 0$; i.e. Voltage across s_2 terminal will be maximum & voltage between s_3 & s_1 terminals will be zero. This is called as *Electrical Zero Position of the synchro transmitter.*

Thus input to this unit is the angular position (θ) & output is the set of three voltages.

Synchro Control Transformer : Construction is same as that of transmitter but the rotor is of cylindrical shape. It is also called as *Synchro Receiver.* Output of the transmitter is appliied to the control transformer. The pair of synchro transmitter and receiver together is called as *Synchro Error Detector..*

With this connection a flux pattern identical to that in synchro-transmitter is established in the air-gap of the control transformer; hence its flux axis will be in the same position as that of the transmitters rotor. Therefore the voltage induced in the control transformer rotor winding is proportional to the cosine of the angle between the two rotors.

$\therefore$ $e(t) = K' [V_{rm} \sin \omega t] \cos \varphi$ ...(4)

Where φ = Angular displacement between the two rotors.

If $\varphi = 90^\circ$ two rotors are at right angles & voltage induced in the control transformer rotor will be zero. This is known as *'Electical Zero Position'* of the control transformer (as shown in the diagram). The phasor representation of the displacement of the rotors of synchro - transmitter and synchro receiver from their electrical zero positions.

φ = Angular displacement between the two rotor axes

$= (90 - \theta) + \alpha$

Put this in equ (4)

$$e(t) = K' V_{rm} \sin \omega t \cos (90 - \theta + \alpha)$$

$$= K' v_r \cos [90 - \theta + \alpha]$$

$$= K' v_r \sin (\theta - \alpha)$$

If deviation $\theta - \alpha$ is small ; $\sin (\theta - \alpha) \approx (\theta - \alpha)$

$\therefore$ $e(t) = K' v_r . (\theta - \alpha)$

Thus $e(t) \infty (\theta - \alpha)$

i.e. $e(t) = K_s (\theta - \alpha)$;

K_s = Sensitivity of the error detector.

Taking L.T.

$$E(s) = K_s [\theta - \alpha]$$

Hence synchro can be used as an error detector giving a voltage at the rotor winding terminals of the transformer as output which is proportional to the difference between the shaft positions of two rotors for small deviations.

TACHOMETER

Tachometer is a rotating device which converts mechanical rotations into proportionate electrical voltage. It has two coils on it's stator namely reference coil and the quadrature coil. These coils are physically placed 90° apart. The rotor is made of a highly conducting material & weighs less so as to have less inertia.

A tachometer can be classified into two types.

(i) AC tachometer and

(ii) DC tachometer. In a DC tachometer, permanent magnet is used in place of external winding.

In both types the output voltage developed is proportional to the magnitude of angular velocity.

$$e_t(t) \propto \frac{d\theta}{dt}$$

$$\therefore \; e_t(t) = K_t \frac{d\theta}{dt}$$

K_t = Tachometer constant in V/rad/sec

$e_t(t)$ = output voltage in volts

& θ = any displacement in radians

To obtain the transfer function

Taking L.T.

$$E_t(s) = K_t \cdot s\,\theta(s) \quad \frac{E_t(s)}{\theta(s)} = K_t \cdot s$$

Applications:

(i) **Speed measurement :** In this application tachometer output is directly connected to a voltmeter which is calibrated in terms of speed.

(2) Position Control System

(3) Speed Control System

The block diagram for below in speed control system is as under.

$$K_m = \frac{K_T}{R_a \cdot f}; \quad f = \frac{K_b \cdot K_T}{R_a} + f_0$$

$$T_m = \frac{J}{f}$$

$$\frac{\theta(s)}{E_{ref}} = \frac{K_A \cdot K_m}{s(1+T_m s) + K_A K_T s}$$

POTENTIOMETER

It is electro mechanical transducer that converts mechanical energy into electrical energy. The input to the device is in the form of a mechanical displacement either linear or rotational & output is the voltage. It is made up of wire wound or conductive plastic resistive elements. A constant voltage (v) is applied across the fixed terminals. The variable arm is connected to the shaft of the system whose position is to be controlled.

For 'N' turns potentiometer the total displacement of the variable arm is $2\pi N$ rad. The output volt e(t) will be proportionate to the shaft position.

$$e(t) \propto \theta(t)$$

$$\therefore \quad e(t) = K_s\,\theta(t) \qquad ...(1)$$

K_s = Proportionality constant

$$= \frac{V}{2\pi N} \text{ volts/rad} \qquad ...(2)$$

K_s is also called as sensitivity of potentiometer.

For comparing the positions of two remotely located shafts, two pots in parallel are connected across the same voltage source.

$$e(t) = e_1(t) - e_2(t) = K_s [\theta_1(t) - \theta_2(t)] \qquad ...(3)$$

Taking L.T. of (1) and (3)

$$\frac{E(s)}{\theta(s)} = K_s \qquad ...(4)$$

$$\frac{E(s)}{[\theta_1(s) - \theta_2(s)]} = K_s \qquad ...(5)$$

Block Diagrams of eqn. (4) and (5) are as shown below.

Applications :

(i) For position control Systems,

(ii) For comparing the position of two remotely located shafts.

MATHEMATICAL MODELLING OF CONTROL SYSTEM

Purpose of Mathematical Modelling

For any physical system we won't be able to get the idea about its behaviour before actual installation unless we analyse it. After analysing we can get the idea about its stability & response. For analysis we require the mathematical model of the system which gives the relation between the input & output in terms of system parameters. Hence mathematical modelling is essential for analysis as well as for design.

Steps for Preparing the Mathematical Model

1. Get the idea about the components used in the physical system,their purpose & purpose of the whole system.
2. Replace all the components by their block diagram showing the input/output relationship [in forward path as well as in the feedback path].
3. Connect all the components sequentially from Input to Output as well as in the feedback path. Connection will be according to their position in actual system. This will be the Block Diagram of the system.
4. Find the relationship between Input & Output [Transfer Function] of the system by B.D. reduction technique or by SFG or directly by

equations. This will be the mathematical model of the given physical system.

TRANSFER FUNCTION (T.F.)

Definition

The transfer function of a linear time - invariant system is defined as the ratio of the laplace transform of the output variable to the laplace transform (L.T.) of the input variable with zero initial conditions.

Properties

(i) It is in 'S' domain.

(ii) It gives idea about 'Poles' & 'Zeros' of the system.

(iii) It is independent of the input.

(iv) If the system has more than one non-loading element then T.F. of all the elements are determined independently & the overall T.F. can be obtained by multiplying all the T.F. If the elements load each other, the overall T.F. is derived by the basic analysis.

(v) The highest power of the complex variable 'S' in the denominator of the transfer function determines the order of the system.

Limitations

(i) It is valid only for linear time - invariant systems.

(ii) It does not take into account the initial conditions.

(iii) It does not give any idea about how to present output is progressing.

BLOCK DIAGRAM (B.D.)

Definition

It is a convenient graphical representa-tion of the transfer function which gives the clearcut idea about the components & signals of the system. It can be used for linear as well as non-linear system. The blocks are connected by lines with arrows representing the unidirectional flow of information from the output of one block to the input of the other. The blocks are of two types:

1. Block representing gain

r = input

c = output

K = gain

$\frac{c}{r} = K$

i.e. c = r . K

Block Diagram of Closed Loop C.S. & its Transfer Function

R(s) = Input signal

C(s) = Output signal

B(s) = Feedback signal

E(s) = Error signal

$E(s) = R(s) \pm B(s)$ ...(1)

In above expression the '+' sign corresponds to positive feedback and '–' sign corresponds to negative feedback.

It should be noted that if nothing is mentioned about the type of feedback, then negative feedback system should be assumed.

Now,

G(s) = Forward path $T.F. = \frac{C(s)}{E(s)}$ ...(2)

H(s) = T.F. of feedback element

$$= \frac{B(s)}{C(s)} \quad ...(3)$$

G(s) . H(s) = Loop T.F. $= \frac{B(s)}{E(s)}$ ...(4)

T(s) = Closed Loop T.F. $= \frac{C(s)}{R(s)}$

For negative feedback system, we have,

E(s) = R(s) – B(s) from (1)

Using (3); E(s) = R(s) – H(s) . C(s) ...(5)

Using (2); C(s) = G(s) . E(s) ...(6)

Using (5) C(s) = G(s) {R(s) – H(s) .C(s)}

C(s) {1 + G(s) . H(s)} = G(s) . R(s)

$$T(s) = \frac{C(s)}{R(s)} = \frac{G(s)}{1 + G(s).H(s)} \quad ...(7)$$

For positive feedback system :

$$T(s) = \frac{C(s)}{R(s)} = \frac{G(s)}{1 - G(s).H(s)} \quad ...(8)$$

Now if we substitute (2) in (5) for C(s)

$$E(s) = R(s) - H(s).\ G(s)\ .\ E(s)$$

$$\therefore\ E(s)\ [1 + G(s)\ .\ H(s)] = R(s) \qquad ...(9)$$

$$E(s) = \frac{R(s)}{1+G(s).H(s)}$$

This expression is for a negative feedback system. For a positive feedback system it will become

$$E(s) = \frac{R(s)}{1-G(s).H(s)}$$

The closed loop C.S. can also be represented by following block diagram (B.D.)

Note : $\left[\frac{C(s)}{R(s)} = T(s) = \frac{G(s)}{1+G(s).H(s)}\right]$

Hence we can conclude that the B.D. of a closed loop system can be reduced to that of open loop system, but T.F. of the block is different. For practical systems B.D. becomes complicated, hence it is necessary to reduce the B.D. in order to get the T.F.

Block-Diagram Reduction Techniques

(1)	Two Block in Series	Equivalent
(2)	Two Blocks in Parallel	Equivalent
(3)	Eliminating parallel paths	Equivalent
(4)	Eliminating feedback path	Equivalent
(5)	Shifting a take-off point ahead of element	Equivalent
(6)	Shifting take-off point before an element	Equivalent
(7)	Moving take-off point ahead of summing point	Equivalent
(8)	Moving take-off point beyond summing point	Equivalent
(9)	Shifting a summing point ahead of element	Equivalent
(10)	Shifting a summing point beyond an element	Equivalent

Block-Diagram & T.F. of Multi Input System

In such case the output is obtained by the principle of super-position.

Let us consider a two input system, wherein

R_1 (s) & R_2 (s) are two Inputs.

Step 1 : Obtain the output C(s) by considering input R_1 (s) alone, assuming R_2 (s) to be zero.

∴ In the forward path G_1 (s) and G_2(s) will be in series.

∴ The equivalent will be $G_1(s) \, . \, G_2(s)$

∴ For this system

$$\frac{C(s)}{R_1(s)} = \frac{G_1(s).G_2(s)}{1+G_1(s).G_2(s).H(s)}$$

Step 2 : Now consider R_2(s) alone and let $R_1(s) = 0$

H(s) & G_1(s) will appear in feedback path both connected in series.

∴ The equivalent of feedback path will be $G_1(s) \, . \, H(s)$

$$\therefore \frac{C(s)}{R_2(s)} = \frac{G_2(s)}{1+G_1(s).G_2(s).H(s)}$$

Step 3 : Now obtain the output by super position theorem.

∴ C(s) = sum of outputs due to both inputs separately.

$$= \frac{G_1(s)\,G_2(s)}{1+G_1(s).G_2(s).H(s)} R_1(s) + \frac{G_2(s)}{1+G_1(s).G_2(s).H(s)} R_2(s)$$

∴ In general,

$$C_i(s) = \sum_{j=1}^{n} G_{ij}(s).R_j(s)$$

Where $R_j(s) = j_{th}$ input

$G_{ij}(s)$ = T.F. between i^{th} output & j^{th} input with all other inputs reduced to zero. $C_i(s) = i^{th}$ output

SIGNAL FLOW GRAPH

Definitions

It is a graphical representation between the variables of a set of linear algebraic equation. It is used to find the transfer function. Transfer function is obtained by Mason's Gain formula after constructing the S.F.G of the system. The nodes represent the variables & signal travels in the direction of arrow along the branches. The signal travelling along the branches is multiplied by the gain of the branch. For complicated C.S. the Block-Diagram reduction technique becomes very tedious & time consuming for getting the T.F. Hence SFG may be useful. It is used for linear C.S. only.

Terms Associated with SFG

(i) *Input Node (Source) :* It is a node which has only outgoing branches.

(ii) *Output Node :* It is a node which has only incoming branches.

(iii) *Path :* It is the traversal of connected branches in the direction of the branch arrows such that no node is traversed more than once.

(iv) *Forward Path :* It is the path from input to output node along which no node is traversed more than once.

(v) *Loop :* It is the path which originates & terminates at the same node.

(vi) *Non-touching loops :* If the loops do not posses any common node, they are non-touching loops.

(vii) *Path gain :* The product of branch gains encountered in traversing a path is called as path gain.

(viii) *Loop gain :* The product of branch gains encountered in traversing a loop is called as loop gain.

Rules for Constructing SFG

(i) The value of variable representing a node is equal to the sum of all the signals entering that node.

(ii) The value of the variable represented by the node is transmit-ted through all the branches leaving the node.

(iii) The parallel branches between two nodes can be replaced by a single branch having gain equal to the sum of the gain of all the parallel branches.

(iv) Similarly for series, the equivalent is the product.

Mason's Gain Formula

$$M = \frac{Y_{out}}{Y_{in}} = \sum_{k=1}^{n} \frac{M_k \cdot \Delta_k}{\Delta}$$

Where

n = Number of forward paths.

M_k = Gain between Y_{out} & Y_{in} along the kth forward path.

Δ = 1 – (Sum of all individual loop gains) + (Sum of gain products of all possible combinations of two non-touching loops) – (Sum of gain products of all possible combination of three non-touching loops) + ...

$\Delta_k = \Delta$ for that path of the SFG which is not touching the kth forward path.

Y_{in} = Input node.

Y_{out} = Output node.

The formula is applicable between input and output node only.

EFFECT OF FEEDBACK ON CONTROL SYSTEM

Effects of Negative (De-generative) Feedback on the system

Sensitivity : It is defined as the ratio of % change in T.F. to the % change in gain G(s)

$$S_G^T = \frac{\%\text{ change in T.F. }[T(s)]}{\%\text{ change in gain }[G(s)]}$$

$$= \frac{\partial T/T(s)}{\partial G/G(s)} = \frac{\partial T}{\partial G} \times \frac{G(s)}{T(s)}$$

For open - loop system T(s) = G(s)

$$\therefore\ S_G^T = 1$$

For closed loop system

$$T(s) = \frac{G(s)}{1+G(s).H(s)}$$

$$\therefore\ \frac{\partial T(s)}{\partial G(s)} = \frac{1+G(s).H(s) - G(s).H(s)}{\{1+G(s).H(s)\}^2}$$

$$\therefore\ S_G^T = \frac{1}{\{1+G(s).H(s)\}^2} \times \frac{G(s)[1+G(s)H(s)]}{G(s)}$$

$$= \frac{1}{1+G(s).H(s)}$$

$\therefore$ Sensitivity of closed loop system is reduced by fac tor $\frac{1}{1+G(s).H(s)}$ due to negative feedback.

Similarly sensitivity due to feedback element is defined as

$$S_H^T = \frac{\partial T}{\partial H} \times \frac{H(s)}{T(s)}$$

$\therefore$ S^T_H for closed loop system

$$= \frac{-G(s)^2}{\{1+G(s).H(s)\}^2} \times \frac{H(s) \times \{1+G(s).H(s)\}}{G(s)}$$

$$= \frac{-G(s).H(s)}{1+G(s).H(s)}$$

For large systems, G(s).H(s)+1G(s) .H(s)

$\therefore$ $S^T_H \approx 1$

It is clear from the equation of sensitivity that changes in H(s) directly affect the output. Hence feedback element should be selected such that it must not change with the environment.

System Dynamics:

Let $G(s) = \frac{k}{s+\alpha}$

$\therefore$ For open loop system; $\frac{C(s)}{R(s)} = \frac{k}{s+\alpha}$

For unit impulse input i.e. for R(s) = 1

$$C(s) = \frac{k}{s+\alpha}$$

To obtain the value of output C in time domain, take laplace transform.

$\therefore$ $c(t) - L\,[C(s)] = k\,e^{-\alpha t}$ ...(1)

For unity feedback system, i.e. for H(s) = 1 & same G(s);

$$T(s) = \frac{C(s)}{R(s)} = \frac{G(s)}{1+G(s).H(s)} = \frac{k}{s+(k+\alpha)}$$

For impulse Input, $C(t) = ke^{-(k+\alpha)t}$ ...(2)

The response for eqn (1) and (2)

It is clear that due to negative feedback response becomes fast.

Effect of Disturbance Signal

Consider an open - loop system.

G_1 & G_2 are two forward path blocks; R is input.

Let 'N' represent the disturbance signal

In open loop mode (i.e. H = 0), if input R is assumed to be zero than the T.F. due to N alone will be

$$\therefore \left.\frac{C}{N}\right|_{R=0} = G_2$$

If system is operating in closing loop mode [dotted line] then

$$\therefore \left.\frac{C}{N}\right|_{R=0} = \frac{G_2}{1+G_1G_2H}$$

If $G_1 G_2 H + 1 \approx G_1G_2 H$

$$\left.\frac{C}{N}\right|_{R=0} = \frac{G_2}{G_1G_2H} = \frac{1}{G_1H}$$

$\therefore$ Effect of disturbance signal will be reduced sufficiently from G_2 to $\frac{G_2}{G_1 G_2 H}$ by negative feedback.

Gain

For open - loop system; Gain = $|G(s)|$

For closed loop system, Gain

$$\left|\frac{G(s)}{1+G(s).H(s)}\right|$$

$\therefore$ Gain reduces.

Advantages of Negative Feedback

(i) Sensitivity due to parameter variation reduces

(ii) System response becomes fast.

(iii) Effect of disturbance signal is less.

Disadvantages of Negative Feedback

(i) Gain is reduced.

(ii) If $G(s) . H(s) = -1$; $T(s)$

$= \infty$. Hence system looses stability.

(iii) Due to introduction of feedback, no of components increases. Hence weight, cost & complexity of the system increases.

Effects of Positive [Regenerative] Feedback on the System

For +ve feedback system;

$$T(s) = \frac{C(s)}{R(s)} = \frac{G(s)}{1-G(s)\,H(s)}$$

$\therefore$ Gain of the system increases, but if $G(s) . H(s) = 1$; $T(s) = \infty$ and system will become unstable.

Physical Significance of Sensitivity

Due to environmental changes, aging of components etc, the value of G(s) changes, hence output also changes. The requirement of the system is that due to all these changes output should not be affected. Hence sensitivity of the output to all these changes it should be minimum. For open – loop system all these variations cause the system output to be inaccurate. Closed - loop system senses these changes in the output & attempts to change the system input so as to reduce the error in the output. Hence sensitivity due to parameter variation reduces.

ANALOGOUS OF MECHANICAL SYSTEMS

Analogous Systems

Systems whose differential equations are of identical form are called as **anologous systems.** The technique of analogous systems is useful for the study of various systems like electrical, mechanical, thermal, pneumatic etc. In general electrical & electronic systems are much easier to deal with experimentally, hence it is convenient to study a non-electrical system in terms of its electrical analogue. Analogies between two systems breakdown when the region of operation is extended too far.

Modelling of Mechanical System

The modelling of mechanical systems can be classified into two groups for two different motions viz. translationl (linear) motion & rotational motion.

(A) *Translational Motion :* It is the motion that takes place along a straight line. The variables used to define translational motion are force, accelerational, velocity & displacement. Following elements are involved in translational motion.

1. **Mass :** It is considered as an indication of the property of an element which stores the kinetic energy of translational motion. It is analogues to inductance of electric network.

$$M = \text{Mass} = \frac{W}{g}$$

W = Weight;

g = Acceleration due to gravity.

a = Acceleration;

v = Velocity

x(t) = Displacement

f(t) = Force acting on the mass

$$= M \cdot a(t) = M \cdot \frac{d\,v(t)}{dt} = M \frac{d^2\,x(t)}{dt^2}$$

2. **Linear Spring :** It is considered to be an element that stores potential energy. It is analogues to capacitor.

$f(t) = K\,x(t)$

K = Stiffness (Constant) of

the spring in N/m.

3. **Friction :** Whenever their is motion between two elements, frictional force exists.

$$f(t) = f_B \frac{d\,x(t)}{dt}$$

f_B = Viscous frictional coefficient in N/m/sec.

(B) *Rotational Motion :* It is defined as a motion about a fixed axis. Variables used to define rotational motion are torque (T), angular velocity (ω), angular displacement (θ) & angular acceleration (α). Following elements are involved in rotational motion.

1. **Inertial (J) :** It is considered as an indication of property of an element which stores the kinetic energy of rotational motion. When torque 'T' is applied to a body with inertia 'J', the torque equation is

$T(t) = J \cdot \alpha\,(t)$

$$= J \frac{d\,\omega\,(t)}{dt} = J \frac{d^2\,\theta(t)}{dt}$$

2. **Torsional Spring :** Torsional Spring constant (K) in torque per unit angular displacement is devised to represent compliance of rod or shaft when it is subjected to an applied torque.

$T(t) = K \cdot \theta(t);$

K is N.m/rad

3. **Frictrion for Rotational Motion :** It is the same as that for translational motion.

$$T(t) = B \cdot \frac{d\,\theta\,(t)}{dt}$$

Where B = is the friction coefficient in N.m/rad/sec.

Mechanical Translational System

Consider a spring - mass - damper system. The dash pot (A device that provides viscous friction) consists of a piston & oil-filled cylinder. Any relative motion between the piston rod & cylinder is resisted by the oil, as the oil must flow around the piston. Dash - pot; hence provides viscous friction, hence absorbs energy. This absorbed energy is dissipated as heat. To obtain the Transfer function following steps should be followed:

(i) Write the differential equation of the system.

(ii) Take the Laplace transform assuming all initial conditions zero.

(iii) Take the ratio of output to Input.

x(t) = Force applied to the Mass → Input

y(t) = Displacement of the Mass → output

Frictional Force of the dash pot $\propto \frac{d\,y(t)}{dt} = f\frac{d\,y(t)}{dt}$

Spring Force $\propto$ y(t) = K y(t)

F = Output force

$$\propto \frac{d^2\,y(t)}{dt^2} = M\frac{d^2\,y(t)}{dt^2} \propto \frac{d^2\,y(t)}{dt^2} = M\frac{d^2\,y(t)}{dt^2}$$

$$\therefore\ F = M\frac{d^2\,y(t)}{dt^2} = x(t) - f\frac{d\,y(t)}{dt} + K\,y(t)$$

$$\therefore\ x(t) = M\frac{d^2\,y(t)}{dt^2} + f\frac{d\,y(t)}{dt} + K\,y(t)$$

Take L.T.

$$\therefore\ X(s) = [M\,s^2 + f\,s + K]\,Y(s)$$

$$\therefore\ \frac{Y(s)}{X(s)} = \frac{1}{Ms^2 + fs + K} \quad ...(1)$$

Mechanical Rotational System

This system consists of a load - inertia (J) & a viscous friction damper (f).

T = Applied torque Input & θ = Angular Displacement (Output)

$$T = J\alpha + f\omega = J\frac{d\omega}{dt} + f\omega$$

[Torsional Spring effect is neglected]

$$\therefore\ T = J\frac{d^2\theta}{dt^2} + f\frac{d\theta}{dt}$$

Take L.T.

$$T(s) = (J\,s^2 + f\,s)\,\theta(s)$$

$$\therefore\ \frac{\theta(s)}{T(s)} = \frac{1}{J\,s^2 + f\,s} \quad ...(2)$$

Electrical System

[A]For series circuit

For this circuit, E_i = Input

$$e_i(t) = R\,i(t) + L\frac{d\,i(t)}{dt} + \frac{1}{C}\int i(t)dt$$

$$i(t) = \frac{dq(t)}{dt}; \quad q = \text{Charge}$$

$$\therefore \; e_i(t) = R.\frac{dq(t)}{dt} + L\frac{d^2q(t)}{dt^2} + \frac{1}{C}q(t)$$

Take L.T.

$$\therefore \; E_i(s) = \left[Ls^2 + Rs + \frac{1}{C}\right]\theta(s)$$

$$\therefore \; \frac{\theta(s)}{E_i(s)} = \frac{1}{Ls^2 + Rs + \frac{1}{C}} \qquad ...(3)$$

[B] For parallel Circuit

For this circuit $i_s(t)$ = Input

$$i_s(t) = \frac{e(t)}{R} + \frac{1}{L}\int e(t)\,dt + C\frac{de(t)}{dt}$$

$$e(t) = \frac{d\psi(t)}{dt}; \quad \psi = \text{Flux - linkage}$$

$$\therefore \; i_s(t) = \frac{1}{R}\frac{d\psi(t)}{dt} + \frac{1}{L}\psi(t) + C\frac{d^2\psi(t)}{dt^2}$$

Take L.T.

$$\therefore \; I_s(s) = \{Cs^2 + \frac{1}{R}s + \frac{1}{L}\}\,\psi(s)$$

$$\therefore \; \frac{\psi(s)}{I(s)} = \frac{1}{Cs^2 + \frac{1}{R}s + \frac{1}{L}} \qquad ...(4)$$

When equation (1) and (2) of Mechanical systems in sections 1.7.3 and 1.7.4 are equated with equation (3) and (4) of Electrical System in section1.7.5 we get the analogous quantities. They are as tabulated below :

GEAR TRAINS, LEVERS & TIMING BELTS

Gear train, lever and timing belt are the mechanical device that transmits energy from one part of a system to another in such a way that force, speed

& displacement are altered. The relationship between Torque (T), angular velocity (ω), angular displacement (θ) and the gear ratio (N_1/N_2) are derived froın the following facts :

(i) Number of teeth on the surface of gear ∞ radius (r_1 & r_2) of gears

$$\therefore \quad N_1 r_2 = N_2 r_1 \qquad ...(1)$$

(ii) The distance travelled along the surface of each gear is same.

$$\therefore \quad \theta_1 r_1 = \theta_2 r_2$$

$$T_1\theta_1 = T_2\theta_2 \qquad ...(2)$$

(iii) If losses are neglected, the work done by both the gears is same

$$T_1\theta_1 = T_2\theta_2 \qquad ..(3)$$

(iv) Work done

$$W_1 = \frac{d\theta_1}{dt} \ \& \ W_2 = \frac{d\theta_2}{dt} \qquad ...(4)$$

Combining (1) to (4) we get

$$\frac{r_1}{r_2} = \frac{N_1}{N_2} = \frac{\theta_2}{\theta_1} = \frac{T_1}{T_2} = \frac{W_2}{W_1} \qquad ...(5)$$

ELECTRICAL ANALOGUE OF GEAR TRAIN

The electrical analogue of such mechanical systems shown in Figure 1.32 is a Transformer. For transformer we have

$$\frac{V_1}{V_2} = \frac{I_2}{I_1} = \frac{N_1}{N_2} \qquad ...(6)$$

$$r_{e1} = \left(\frac{N_1}{N_2}\right)^2 r_{e2} \quad x_{e1} = \left(\frac{N_1}{N_2}\right)^2 x_{e2} \qquad ...(7)$$

Where

r_{e1} = Equivalent resistance of trans-former referred to primary side

r_{e2} = Equivalent resistance of trans-former referred to secondary side.

The general torque equation is

$$T = J\alpha + f\omega = J\frac{d^2\theta}{dt^2} + f\frac{d\theta}{dt}$$

$$\therefore \quad T_1 = J_1\frac{d^2\theta_1}{dt^2} + f_1\frac{d\theta_1}{dt} \qquad ...(8)$$

$$T_2 = J_2\frac{d^2\theta_2}{dt^2} + f_2\frac{d\theta_2}{dt} \qquad ...(9)$$

From equation (5)

$$T_1 = \left(\frac{N_1}{N_2}\right) T_2 \text{ and } \theta_2 = \left(\frac{N_1}{N_2}\right) \theta_1$$

Substituting the value of T_2 from eq (9)

$$T_1 = \left(\frac{N_1}{N_2}\right) J_2 \left\{\frac{N_1}{N_2}\cdot\frac{d^2\theta_1}{dt^2}\right\} + \left(\frac{N_1}{N_2}\right) f_2$$

$$\left\{\frac{N_1}{N_2}\cdot\frac{d\theta_1}{dt}\right\}$$

$$\therefore\ T_1 = \left(\frac{N_1}{N_2}\right)^2 J_2 \cdot \frac{d^2\theta_1}{dt^2} + \left(\frac{N_1}{N_2}\right)^2 f_2 \frac{d\theta_1}{dt} \quad \text{...(10)}$$

Equating (8) and (10)

$$\therefore J_1 = \left(\frac{N_1}{N_2}\right)^2 J_2 \ \& \ f_1 = \left(\frac{N_1}{N_2}\right)^2 f_2 \quad \text{...(11)}$$

These are the quantities obtained when reflecting from gear 2 to gear 1. Electrical equivalent of eqn (11) is eqn (7).

Significance

For experimental purpose the mechani-cal elements & variables can be replaced by their electrical analogues. In the working range the result can be transferred to mechanical system as experimentation is easy for electrical systems.

CONTROL SYSTEM COMPONENTS

These are the components/devices which are used as controllers/sensors for the plant. They are connected in forward as well as in feed-back paths.

In control system analysis they are replaced by their mathematical equivalent i.e. by their transfer function. Let us discuss some important control system components.

SERVO MOTORS

Some sort of drive is required to actuate the various mechanical components in control systems.

The motors which are used for such low power applications are called as servo motors. They are of two types D.C. as well as A.C. Servo - motors.

Armature Control :

For this control, e_f & i_f are constant.

The armature voltage 'V_a' will be the input and displacement θ will be output.

$T_m \propto \Phi_f \cdot i_a;$

$\Phi_f \propto i_f \rightarrow$ Constant

$$\therefore T_m = K_T i_a \quad ...(1)$$

K_T = Motor torque constant

$e_b \propto N$; N = Speed in rpm

$$\therefore e_b \propto \frac{d\theta}{dt}$$

$$e_b = K_b \cdot \frac{d\theta}{dt} \quad ...(2)$$

K_b = Back emf constant

For armature circuit,

$$V_a = R_a i_a + L_a \frac{d i_a}{dt} + e_b \quad ...(2)$$

& Torque equation is,

$$T_m = J \frac{d^2\theta}{dt^2} + f_0 \frac{d\theta}{dt} \quad ...(3)$$

Take L.T. of equations (6 to 9)

$$\left.\begin{aligned} T_m(s) &= K_T I_a(s) \\ E_b(s) &= K_b s.\theta(s) \\ [V_a(s) - E_b(s)] &= [R_a + sL_a]\, I_a(s) \\ T_m(s) &= s[Js + f_0]\theta(s) \end{aligned}\right\} \quad ...(5)$$

Using equation (10) we can draw the B.D. well as S.F.G; and we can find the T.F. $\frac{\theta(s)}{V_a(s)}$.

Note : In equation (10) we can add one more equation in following way

$T_m(s) = (J s + f_0)\, \omega_0(s);\ \theta(s) = s\, \theta(s) = \omega(s)$

$\therefore$ B.D. becomes

Transfer Function :

$$\frac{\theta(s)}{V_a(s) - E_b(s)} = G(s) = \frac{K_T}{(R_a + sL_a)(f_0 + sJ)} \quad H(s) = K_b$$

$$\frac{\theta(s)}{V_a(s)} = \frac{G(s)}{1+G(s)H(s)}$$

$$= \frac{K_T/(R_a+sL_a)(f_0+sJ)}{1+\dfrac{K_T.K_b}{(R_a+sL_a)(f_0+sJ)}}$$

$$\therefore \frac{\theta(s)}{V_a(s)} = \frac{K_T}{(R_a+sL_a)(f_0+sJ)+K_T K_b}$$

$$\therefore \text{T.F.} = \frac{\theta(s)}{V_a(s)}$$

$$= \frac{K_T}{s[K_T.K_b+(R_a+sL_a)(f_0+Js)]}$$

L_a is very small; hence can be neglected.

$$\therefore \frac{\theta(s)}{V_a(s)} = \frac{K_T/R_a}{s\left[\left(\dfrac{K_T.K_b}{R_a}+f_0\right)+Js\right]}$$

$$= \frac{K_T/R_a}{s[f+Js]}$$

Where

$$f = \frac{K_T.K_b}{R_a}+f_0$$

= Effective viscous coefficient

[$f > f_0$; this indicates that back emf effectively increases the viscous friction]

$$\therefore \frac{\theta(s)}{V_a(s)} = \frac{K_T/R_a f}{s\left(\dfrac{J}{f}s+1\right)} \quad \text{...(i)}$$

$$\frac{\theta(s)}{V_a(s)} = \frac{K_m}{s(1+\tau_m s)} \quad \text{...(ii)}$$

Where $K_m = \dfrac{K_T}{R_a.f}$ = Motor gain &

$\dfrac{J}{f} = \tau_m$ = Motor time constant.

Applications :

This method is employed for control of large servo motors, especially because of the existence of back emf in the control circuit which contibutes like that of additional friction, thus providing additional damping.

The drawback is that large power amplification is required.

Whenever there is use of D.C. servo-motor in the system, for mathematical analysis purpose it is replaced by equation (5) or (11) depending on whether it is field control or armature control.

A.C. Servo-Motor

Constructionally this motor has two windings on the stator and a rotor similar to that of a induction motor.

It differs with normal induction motors in two ways :

Rotor circuit has high resistance i.e. it has low $\frac{X}{R}$ ratio, hence torque (T) verses speed (N or ω) Characteristic is linear.

- Two seperate voltages are connected across the two stator windings. The voltages across these windings i.e. the reference winding (v_r) and the control winding (e_c) are such that they have a 90° electrical phase displacement between them. Because of this phase displacement a rotating magnetic field gets developed in the air gap which produces a torque on the rotor and rotates it.

Principle :

Servo motor is basically used to convert the control signal or voltage into proportionate angular displacement. In a AC servo motor the voltage connected across the reference winding is fixed at a desired constant value ; whereas the voltage connected control winding (e_c) is the control voltage which may vary in phase as well as in magnitude.

Due to this reason the torque developed is proportional to control voltage.

$$T \propto e_c.$$

Salient Features :

(i) The starting torque of servo - motor is proportional to the rms value of control voltage.

$\therefore T_{st} \propto e_c(t).$

$\therefore$ T - N characteristic For various e_c (t) has negative slope.

(ii) When $e_c = e_{c0} = 0,$

The curve passes through the origin & motor develops a decelerating torque.

(iii) The torque developed is a function of control voltage as well as the angular velocity.

$\therefore T=f(e_c, \theta)$

(iv) As the servo - motor never operates at high speed, curve can be linearized about the operating point.

(v) $$T.F. = \frac{\theta(s)}{E_c(s)}$$

$$= \frac{K_m}{s(1+\tau_m s)}$$

Where $K_m = \frac{K}{f+f_0}$

= Motor gain constant

$\tau_m = \frac{J}{f+f_0}$ = Motor time constant

K & f are evaluted from No load & Blocked rotor (stall torque) tests at rated voltage.

$$K = \frac{\text{Stall torque at rated voltage}}{\text{Rated control phase voltage}} \quad \&$$

$$f = \frac{\text{Stall torque at } e_c \text{ rated}}{\text{No load speed at } e_c \text{ rated}}$$

SOLVED EXAMPLES

Example 1:

A motor generator set is shown in Figure. The system is used as a speed control system. D.C. Motor has a constant field excitation & drives an inertia & friction load. D.C. generator supplies armature voltage & is driven at constant speed to give kg volts per field amp. Using K_T & K_t as motor & tacho constant respectively.

(i) Draw the B.D. indicating T.F. of each block. (ii) Derive the T.F. $\frac{\omega(s)}{E_i(s)}$

Solution:

(a) The motor 'M' is driving a load. It is operating in armature controlled mode. The speed is sensed by the tachogenerator 'T' & proportionate voltage is fed-back. E_i is the reference Input voltage to the generator field (after amplification). Generator voltage is controlling the field current. Hence the components are :

(i) Motor (ii) Generator (iii) Load (iv) Tachometer

(v) Amplifier (vi) Potentiometer

The B.D. is as under : E_g = Generated emf E_b = Back emf

(b) Transfer Function

Step 1 : Solve the internal feedback loop :

$$G(s) = \frac{1}{(L_m + L_q)s + (R_m + R_g)} \times K_T \times \frac{1}{J.s + f}$$

& $\quad H(s) = K_b$

$$\therefore \frac{\omega(s)}{E_g(s)} = \frac{G(s)}{1 + G(s).H(s)}$$

$$= \frac{K_T}{(L_m + L_g)J s^2 + [(R_m + R_g)J + f(L_m + L_g)]s + f(R_m + R_g)}$$

Step 2 : Solve the external feedback loop:

For this loop

$$G(s) = \frac{K_A . K_g . K_T}{\{J(L_m + L_q)s^2 + [J(R_m + R_q) + f(L_m + L_q)]s + f(R_m + R_q)\}(R_f + sL_f)}$$

& $H(s) = K_t$

$$\therefore \frac{\omega(s)}{E_i(s)} =$$

$$\frac{K_A . K_g . K_T}{J.L_f(L_m + L_g)s^3 + [J(R_m + R_g) + f(L_m + L_g)].L_f + JR_f(L_m + L_g)]s^2 + [L_f(R_m + R_q) + R_f\{J(R_m + R_g) + f(L_m + L_q)\}]s + [(R_m + R_g)R_f + K_A . K_g K_T . K_t]}$$

Example 2:

Simplify the block diagram shown in Fig below and obtain the closed loop transfer function C(s)/R(s)

Solution:

$$\frac{C(s)}{R(s)} = \frac{G_1 G_3 [G_2 + H_1]}{1 + G_2 H_2 + 2G_3 H_3 (G_2 + H_1)}$$

Example 3:

Determine the overall transfer function of the system shown in Figure below :

Solution:

$$\frac{C(s)}{R(s)} = \frac{G_1 G_2 G_3 G_4}{(1+H_2G_3)(1+G_1G_2H_1+G_2H_3)}$$

Example 4:

Obtain the transfer function Y(s)/X(s) of the SFG shown in figure below:

Solution:

$$\frac{Y(s)}{X(s)} + \frac{b_1s^2+b_2s+b_3}{s^3+a_1s^2+a_2s+a_3}$$

Example 5:

For a position control system using d.c. servomotor in the armature control mode shown in Figure below, assume that gain constant K is 10, α is 2, and feedback parameter β is 1, evaluate the sensitivities $S^T{}_K$, $S^T{}_\alpha$ and $S^T{}_\beta$.

Solution:

$$S_K^T = \frac{s^2+2s}{s^2+2s+10} \; ; \; S_\alpha^T = \frac{-2s}{s^2+2s+10}$$

$$S_\beta^T = \frac{-10}{s^2+2s+10}$$

Example 5:

Find the magnitude of sensitivity $\left|S_K^T\right|$ at ω = 5 for the systems shown in Figure below.

Solution:

$\left|S_K^T\right|$ = 0.5 for single loop configuration

$\left|S_K^T\right|$ = 1.0 for two loop configuration.

Example 6:

For the system represented by the following B.D. Find the T.F. $\frac{C(s)}{R(s)}$

Solution:

Let us solve the internal forward path first.

Step 1 : G_3 & G_2 are in parallel, wherein G_3 has (–) sign. G_1 is in series with this parallel combination. Hence the equivalent will be $G_1 (G_2 - G_3)$.

The B.D. reduces to

Step 2 : At error - detector E_1, R(s) is directly coming while G_4 & H_2 are coming in series. Also at E_1, H_1 & H_2 are coming in series. Hence B.D. reduces as shown :

Step 3 : In the forward path enclosed by the dotted lines; 1 & $G_4 H_2$ are in parallel; $G_4 H_2$ has –ve sign and further has substractive nature at the summing point.

Hence its equivalent will be $(1 + G_4H_2)$

There is only one feedback loop. Its equivalent is;

$$\frac{G_1(G_2 - G_3)}{1 + G_1.(G_2 - G_3)H_1 H_2}$$

Hence B.D. Becomes :

Step 4 : $\frac{C(s)}{R(s)}$ will be series combination of these two blocks

$$\frac{C(s)}{R(s)} = \frac{G_1(G_2 - G_3)(1 + G_4 H_2)}{1 + G_1(G_2 - G_3)H_1 H_2}$$

Example 7:

Find the output C(s) for the system represented by the following B.D.

Solution:

In this example there are two input $R_1(s)$ & $R_2(s)$. Hence the output will be obtained by principle of superposition.

[A] Consider $R_1(s)$ only. Hence $R_2(s) = 0$. Therefore B.D. will be

Step 1 : Shift H_3 ahead of G_3.

Step 2 : Solve the inner most feedback loop. Its equivalent will be $\frac{G_3}{1 + G_3 H_2}$ & B.D. becomes;

Step 3 : Solve the feedback loop enclosed in the dotted lines. The equation will be

$$\frac{G_2 G_3 / 1+G_3H_2}{1+\dfrac{G_2G_3 . H_3}{(1+G_3 H_2)G_3}} = \frac{G_2 G_3}{1+G_3 H_2 + G_2 H_3}$$

The B.D. will be;

Step 4 : Solve the last feedback loop

$$\therefore \frac{C(s)}{R_1(s)} = \frac{G_1G_2 G_3 / 1+G_3H_2 + G_2H_3}{1+\dfrac{G_1G_2G_3 . H_1}{(1+G_3 H_2 + G_2H_3)}}$$

$$\therefore \quad C(s) = \frac{G_1 . G_2 . G_3}{1+G_3H_2 + G_2H_3 + G_1G_2G_3H_1} \times R_1(s)$$

[B] Now consider $R_2(s)$ alone; hence $R_1(s) = 0$. Therefore B.D. will be

Step 1 : In the outer feedback path G_1 & $-H_1$ are in series.

Hence their equation will be $-G_1H_1$

Shift the output of H_2 from E_2 to E_1

∴ B.D. becomes

Step 2 : Solve the enclosed feedback loop. Equation will be; $\dfrac{G_2}{1+G_2H_3}$

Also the feedback paths with $-H_1 G_1$ & H_2/G_2 are in parallel.

∴ Equation will be $\left\{-\dfrac{H_2}{G_2} - H_1 G_1\right\}$

The B.D. becomes

Step 3 : There are two feedback blocks, connected in series. Hence the B.D. will be;

Note that it is a positive feedback system; but gain of feedback element is –ve. Hence its resultant effect will be that of negative feedback system.

Step 4 : Solve the feedback loop.

$$\therefore \frac{C(s)}{R_2(s)} = \frac{G_3}{1+\dfrac{G_2G_3}{1+G_2H_3}\left(\dfrac{H_2}{G_2} + H_1G_1\right)}$$

$$= \frac{G_3 . (1+G_2H_3)}{1+G_2H_3 + G_2G_3H_2 + G_1G_2H_1G_3}$$

$$C(s)=\frac{G_3\left(1+G_2H_3\right)}{1+G_2H_3+G_3H_2+G_1G_2H_1G_3}\times R_2(s)$$

[C] ∴ The resultant output will be

C(s) = sum of outputs due to each individual input

$$C(s)=\frac{G_1G_2G_3}{1+G_3H_2+G_2H_3+G_1G_2G_3H_1}R_1(s)$$

$$+G_3\frac{(1+G_2H_3)}{1+G_2H_3+G_3H_2+G_1G_2H_1G_3}\times R_2(s)$$

Example 8:

Draw the B.D. for the R-L-C series circuit

Solution:

For such problems first write the equations for the circuit & from those equations draw the B.D.

For output $e_0(t)=\frac{1}{C}\int i(t)\,dt$

Take L.T.; $E_0(s)=\frac{1}{sC}I(s)$ ...(1)

∴ B.D. for this part will be

For Input

$$e_i(t)=Ri(t)+L\frac{di(t)}{dt}+\frac{1}{C}\int i(t)dt$$

$$=R_i(t)+L\frac{di(t)}{dt}+e_0(t)$$

Take L.T.; $E_i(s) = R\ I(s) + L.s.\ I(s) + E_0(s)$

i.e. $E_i(s) - E_0(s) - R\ I(s) = L\ s\ I(s)$

i.e. $\frac{1}{L}E_i(s)-\left[\frac{1}{L}E_0(s)+\frac{R}{L}I(s)\right]=s\,I(s)$ ...(2)

This equation indicates that s I(s) is the output of a summing device which has three inputs. [L.H.S. of the equation]

Also $I(s)=\frac{1}{s}\times s\,I(s)$ ...(3)

For equation (3) the B.D. will be

∴ The final B.D. can be obtained by equation (1) (2) and (3)

Example 9:

A system is represented by following linear algebric equations. Draw the SFG & find the T.F., $\frac{Y_5}{Y_1}$

$Y_2 = a_{12}\ Y_1 + a_{32};\ Y_3 = a_{23}\ Y_2 + a_{43}\ Y_4;$

$Y_4 = a_{24}\ Y_2 + a_{34}\ Y_3 + a_{44}\ Y_4\ Y_5 = a_{25}\ Y_2 + a_{45}\ Y_4$

Solution:

In SFG, variables are represented by the nodes & the gains are marked on the paths joining the nodes. In the present case there are total 5 variables Y_1, Y_2, Y_3, Y_4 & Y_5

Hence total nodes = 5

Now consider variable Y_2. To this variable informations are coming from variable Y_1 & Y_3 & in the process they are getting multiplied by gains a_{12} & a_{32} respectively. SFG for this equation will be as under.

Note that from Y_3 information is fed-back to Y_1.

By using all the equations we get the SFG as constructed.

Here Y_1 is the input node.

Y_2 is the output node

Y_2 to Y_3 & back to Y_2 is 1st loop.

Y_3 to Y_4 & back to Y_3 is 2nd loop.

Y_4 to Y_4 is the 3rd loop

$Y_2 - Y_4 - Y_3 - Y_2$ is the 4th loop.

$Y_2 - Y_3 - Y_4 - Y_3 - Y_2$ can not be considered as a loop as Y_3 node is traversed more than once.

1st & 2nd loop are touching each other; similarly 2nd and 3rd loop are touching each other; but 1st and 3rd loop are not touching each other as they are don't have any common node.

There are total three forward paths :

1st : $Y_1 - Y_2 - Y_3 - Y_4$ & Y_5

2nd : $Y_1 - Y_2 - Y_4 - Y_5$

3rd : $Y_1 - Y_2 - Y_5$

Gain : $\frac{Y_5}{Y_1}$;

$$M = \frac{Y_5}{Y_1} = \sum_{k=1}^{n} \frac{M_k . \Delta_k}{\Delta};$$

n = no. of forward paths = 3

Δ = 1 – (sum of all individual loop gains) + (sum of gain product of all two non-touching loops)

$= 1 - \{(a_{23} \times a_{32}) + a_{34} \times a_{43}) + (a_{44}) + (a_{24} \times a_{43} \times a_{32})\} + \{a_{23} \times a_{32} \times a_{44}\}$

$M_1 = a_{12} \times a_{23} \times a_{34} \times a_{45};$

$\Delta_1 = 1$ {As all the loops are touching this forward path}

$M_2 = a_{12} \times a_{24} \times a_{45};\ \Delta_2 = 1$

{As all the loops are touching this forward path}

$M_3 = a_{12} \times a_{25};$

$\Delta_3 = 1 - \{(a_{34} \times a_{43}) + a_{44}\}$

$$\therefore\ M = \frac{Y_5}{Y_1}$$

$$= \frac{(a_{12} . a_{23} . a_{34} . a_{45}) + (a_{12} . a_{24}\ a_{45}) + a_{12}\ a_{25}\ \{1 - (a_{44} + a_{34}\ a_{43})\}}{1 - \{a_{23}\ a_{32} + a_{34}\ a_{43} + a_{44} + a_{24}\ a_{43}\ a_{32}\} + \{a_{23}\ a_{32}\ a_{44}\}}$$

Note : From the SFG if we are supposed to generate the equation for any variable, then use Rule - 1. e.g. Find the equation for variable Y_2.

According to Rule - 1, Y_2 will be equal to sum of all the signals entering the node. At Y_2 signals are coming from Y_1 & Y_3 & in the process they are multiplied by the gains a_{12} & a_{32} respectively.

$\therefore\ Y_2 = a_{12}\ Y_1 + Y_{32}\ Y_3$

Example 10:

A C.S is represented by the following SFG. Find the T.F. by Mason's Gain Formula.

Solution:

Step 1 : Find out all possible loops & their loop gains.

1st : $G_4 \times (-H_1)$

2nd : $G_5 \times (-H_2)$

3rd : $G_1 \times (-H_2) \times G_8 \times (-H_1)$

Step 2 : Find out all possible combination of two non-touching loops : 3rd loop has common nodes with 1st & 2nd loop, whereas 1st & 2nd loop donot have common node.

∴ They are non-touching

Step 3 : Find out Δ.

$\Delta = 1 - [-G_4H_1 - G_5H_2 + G_1H_2G_8H_1] + [G_4H_1G_5H_2]$

Step 4 : Find out all possible forward paths from R to C.

1st : $G_2G_4G_6$; $M_1 = G_2G_4G_6$;

$\Delta_1 = 1 - [G_5H_2] = 1 + G_5H_2$.

[All other loops are touching this forward path]

2nd : $G_3G_5G_7$; $M_2 = G_3G_5G_7$;

$\Delta_2 = 1 - [-G_4H_1] = 1 + G_4H_1$.

3rd : $G_2G_1G_7$; $M_3 = G_1G_2G_7$ $\Delta_3 = 1$ [All the loops are touching this forward path]

4th : $G_3G_8G_6$;

$M_4 = G_3G_6G_8$ $\Delta_4 = 1$

5th : $-G_2G_1H_2G_8G_6$;

$M_5 = -G_1G_2G_6G_8H_2$ $\Delta_5 = 1$

6th : $-G_3G_8H_1G_1G_7$;

$M_6 = -G_1G_3G_7G_8H_1$ $\Delta_6 = 1$

Step 5 : Find the gain

$$M = \frac{C}{R} = \frac{\sum_{k=1}^{6} M_k \Delta_k}{\Delta}$$

$$\therefore \frac{C}{R} = \frac{\{G_2G_4G_6(1+G_5H_2)+G_3G_5G_7(1+G_4H_1)+G_1G_2G_7 + G_3G_6G_8 - G_1G_2G_6G_8H_2 - G_1G_3G_7G_8H_1\}}{1-[-G_4H_1 - G_5H_2 + G_1H_2G_8H_1]+[G_4H_1G_5H_2]}$$

Example 11:

Find the output of the system whose SFG is as shown :

Solution:

In this case there are total three Inputs. Using Super-position principle.

[A] Consider R_1 alone : R_2 & $R_3 = 0$

∴ SFG will be

Step 1 : Find the loop

1st : $G_1G_2H_2H_1$ → It is the only loop.

Step 2 : Find all possible two non-touching loops.

As only one loop is present, there is no possibility of non-touching loops.

Step 3 : Find Δ

$\Delta = 1 - G_1 G_2 H_1 H_2$

Step 4 : Find all the forward paths & their Δ

1st : G_1G_2

∴ $M_1 = G_1G_2 \; \Delta_1 = 1$

Only one forward path is present.

Step 5 : Find the gain $\frac{C}{R_1}$

$$\therefore \frac{C}{R_1} = \frac{G_1 G_2}{1 - G_1 G_2 H_1 H_2}$$

$$\therefore C = \frac{G_1 G_2}{1 - G_1 G_2 H_1 H_2} \times R_1$$

[B] Consider R_2 alone :

R_1 & $R_3 = 0$

∴ SFG will be

Step 1 : 1st Loop $G_1G_2H_1H_2$.

Step 2 : No two non touching loop available

Step 3 : $\Delta = 1 - G_1 G_2 H_1 H_2$.

Step 4 : Only one forward path

$M_1 = G_2$ & $\Delta_1 = 1$

Step 5 : $\text{Gain} = \frac{C}{R_2} = \frac{G_2}{1 - G_1G_2H_1H_2}$

$$\therefore\ C = \frac{G_2}{1-G_1G_2H_1H_2} \times R_2$$

[C] Consider R_3 alone : R_1 & R_2 = 0 & SFG will be

Step 1 : Only one loop : $G_1G_2H_1H_2$.

Step 2 : No two non-touching loops.

Step 3 : $\Delta = 1 - G_1G_2H_1H_2$

Step 4 : Only one forward path :

$M_1 = H_1G_1G_2$ $\Delta_1 = 1$

Step 5 : $\text{Gain} = \dfrac{C}{R_3} = \dfrac{H_1G_1G_2}{1-G_1G_2H_1H_2}$

$$\therefore\ C = \frac{G_1G_2H_1}{1-G_1G_2H_1H_2} \times R_3$$

Finally find the total output by superposition.

C = sum of outputs due to each individual input.

$$\therefore\ C = \frac{G_1G_2 \times R_1 + G_2 \times R_2 + G_1G_2H_1 \times R_3}{1-G_1G_2H_1H_2}$$

Example 12:

Draw the SFG find the T.F.; $\dfrac{E_0}{E_1}$ for the following R-L-C series circuit.

Solution:

For such problems first write the equations for the circuit & from those equation draw the SFG.

Step 1 : Write the equations for the circuit

For Output; $e_0(t) = \dfrac{1}{C}\int i(t)\,dt$ ...(1)

For Input ;

$$e_i(t) = Ri(t) + L\frac{di(t)}{dt} + \frac{1}{C}\int i(t)\,dt$$

Using (1)

$$e_i(t) = R_i(t) + L\frac{di(t)}{dt} + e_0(t) \qquad \text{...(2)}$$

Now equation (1) & (2) are differential equations.

$\therefore$ SFG can not be constructed.

Step 2 : Transfer equation (1) & (2) in S-domain.

$\therefore$ Take Laplace Transform

$$\therefore\ E_0(s) = \frac{1}{sC}\ I(s) + \text{i.e. } s\ E_0(s) = \frac{1}{C}\ I(s) \quad ...(3)$$

$$\&\ E_i(s) = R\ I(s) + s\ L\ I(s) + E_0(s)$$

$$\text{i.e. } S\ I(s) = \frac{1}{L}\ E_i(s) - \frac{R}{L}\ I(s) - \frac{1}{L}\ E_0(s) \quad ...(4)$$

With equation (3) and (4) we can draw the SFG.

Step 3 : Decide the nodes :

(i) Input Node : $E_i(s)$

(ii) Output Node : $E_0(s)$

Other nodes : From equation (3),

(iii) Node s $E_0(s)$:

(iv) Node I(s) ; from equation (4) &

(v) Node s I(s) from equation (4)

Step 4 : Gains are noted from the equations.

Now draw the SFG using equation (3) and (4)

Note : Here we have created one additional node $E_0(s)$ which has appeared twice. The reason is according to definition of output node; it is the node which has only incoming elements. In the first $E_0(s)$ node, one out going element is also there. Hence this can not be the output node. Hence second $E_0(s)$ node (as output node) is created.

Step 5 : Find T.F.; $\frac{E_0(s)}{E_i(s)}$ by Mason's gain formula

(i) Two loops : s I(s) – I(s) – s I(s)

& s I(s) – I(s) – s$E_0(s)$ – $E_0(s)$ – s I(s)

Loop gains : $\frac{1}{s} \times \frac{-R}{L}$ & $\frac{1}{s} \times \frac{1}{C} \times \frac{1}{s} \times -\frac{1}{L}$

(ii) No two non-touching loops exists.

(iii) $\Delta = 1 - \left[\frac{-R}{sL} + \frac{-1}{s^2 LC}\right]$

(iv) Only one forward path is present i.e.

$$M_1 = \frac{1}{L} \times \frac{1}{s} \times \frac{1}{C} \times \frac{1}{s} \times 1 = \frac{1}{s^2 LC} \quad \& \ \Delta_1 = 1$$

(v) $$\text{T.F.} = \frac{E_0(s)}{E_i(s)} = \frac{1/s^2 LC}{1 + \frac{R}{sL} + \frac{1}{s^2 LC}}$$

$$= \frac{1/LC}{s^2 + \frac{R}{L}s + \frac{1}{LC}}$$

$$\therefore \frac{E_o(s)}{E_i(s)} = \frac{1/LC}{s^2 + \frac{R}{L}s + \frac{1}{LC}}$$

Example 13:

Draw the SFG of a closed loop C.S. & find the T.F.

Solution:

The BD of a closed loop C.S. is

SFG : Identifying the nodes

Nodes : (i) Input Node R(s);

(ii) Output Node ; C(s)

(iii) Create the node after sum device, say E(s)

We have created second C(s) node in order to make it a output node; as from the first C(s) node feedback is coming out.

T.F. (i) Loop : E(s) – C(s) – E(s); Loop Gain = – G(s) H(s)

(ii) Two non-touching loops : NIL

(iii) Forward path : Only one ; $M_1 = 1 \times G(s) \times 1 = G(s)$; $\Delta_1 = 1$

(iv) $\Delta = 1 - (-G(s) . H(s)) = 1 + G(s) . H(s)$

(v) $$\frac{C(s)}{R(s)} = \frac{M_1 . \Delta_1}{\Delta} = \frac{G(s)}{1 + G(s). H(s)}$$

Example 14:

For the system of Example1. 5 draw SFG & find T.F. by Mason's Gain formula.

Solution:

Redraw the given system

Step 1 : Create the nodes after every summing point; say $y_1 y_2 y_3$ & y_4

Step 2 : Equations of the nodes are

$y_1 = R + H_2 y_3$; $y_2 = G_1 y_1 - H_2 y_3 - H_1 y_4$; $y_3 = G_2 y_2$; $y_4 = G_3 y_3$; $c = y_4 + G_4 \cdot R$

Step 3 : Join the nodes as per their equation

Now for T.F. :

(1) Loops and their loop-gains

(i) $y_1 - y_2 - y_3 - y_1$...... $G_1 G_2 H_2$

(ii) $y_2 - y_3 - y_2$...... $- G_2 H_2$

(iii) $y_2 - y_3 - y_4 - y_2$..... $- G_2 G_3 H_1$

(2) Two non-touching loops : NIL

(3) $\Delta = 1 - [G_1 G_2 H_2 - G_2 H_2 - G_2 G_3 H_1]$

(4) Forward path
Path gain M_k Δ_k.

(i) $R - y_1 - y_2 - y_3 - y_4 - C$; $M_1 = G_1 G_2 G_3$ $\Delta_1 = 1$

(ii) $R - C$; $M_2 = G_4$.

$\Delta_2 = 1 - [G_1 G_2 H_2 - G_2 H_2 - G_2 G_3 H_1]$ $= \Delta$.

(5) $$\frac{C}{R} = \frac{M_1 \Delta_1 + M_2 \Delta_2}{\Delta}$$

$$\therefore \frac{C}{R} = G_4 + \frac{G_1 G_2 G_3}{1 + G_2 H_2 + G_2 G_3 H_1 - G_1 G_2 H_2}$$

Example 15:

For the system represented by the B.D. of Example 1.6 draw the SFG & find the T.F.

Solution:

Step 1 : Create the nodes after summing devices & take off points. They are y_1, y_2, y_3 & y_4.

Step 2 : Write the equation for the nodes

$y_1 = R - C$; $y_3 = G_1 y_2 - H_2 C$;

$y_2 = y_1 - H_1 y_4$; $y_4 = G_2 y_3$ C = output node $= G_4 y_3 + G_3 y_4$

Step 3 : Join the nodes

Step 4 : T.F.

(1) Loop
Loop-Gain

(i) $y_1 - y_2 - y_3 - y_4 - C - y_1$: $- G_1G_2G_3$

(ii) $y_2 - y_3 - y_4 - y_2$: $- G_1G_2H_1$

(iii) $y_3 - y_4 - C - y_3$: $- G_2G_3H_2$

(iv) $y_3 - C - y_3$: $- G_4H_2$

(v) $y_1 - y_2 - y_3 - C - y_1$: $- G_1G_4$

(2) Two non-touching loops : NIL

(3) $\Delta = 1 + [G_1G_2G_3 + G_1G_2H_1 + G_2G_3H_2 + G_4H_2 + G_1G_4]$

(4) Forward Path

	Path gain M_k	Δ_k.
(i) $R - y_1 - y_2 - y_3 - y_4 - c$	$G_1G_2G_3$	1
(ii) $R - y_1 - y_2 - y_3 - c$	G_1G_4	1

(5) Transfer Function $\frac{C}{R} = \frac{M_1\Delta_1 + M_2\Delta_2}{\Delta}$

$$\therefore \frac{C}{R} = \frac{G_1(G_2G_3 + G_4)}{1 + [G_1G_2G_3 + G_1G_2H_1 + G_2G_3H_2 + G_4H_2 + G_1G_4]}$$

Example 16:

For the system represented by the B.D. of Ex. 1.7 draw SFG & Find the T.F.

Solution:

Step 1 : Create the nodes. They are R, y_1, y_2, y_3 and C

Step 2 : Equation for Nodes :

$y_1 = R - H_2y_3$; $y_2 = G_1y_1$;

$y_3 = H_1 C - G_4R$; $C = \text{Output} = (G_2 - G_3)\, y_2$.

Step 3 : Join the nodes as per the equation

T.F. (1) Loop
Loop Gain

(i) $y_1 - y_2 - C - y_3 - y_1$ $- G_1G_2H_1H_2$

(ii) $y_1 - y_2 - C - y_3 - y_1$ $- G_1G_3H_1H_2$

(2) Non touching loops : NIL

(3) $\Delta = 1 + G_1 H_1 H_2 (G_2 - G_3)$

(4) Forward Path

Path	Gain M_k	Δ_k.
(i) $R-y_1-y_2-C$	G_1G_2	1
(ii) $R-y_1-y_2-C$	$-G_1G_3$	1
(iii) $R-y_3-y_1-y_2-C$	$G_4H_2G_1G_2$	1
(iv) $R-y_3-y_1-y_2-C$	$-G_4H_2G_1G_3$	1

(5) $$\frac{C}{R}=\frac{M_1\Delta_1+M_2\Delta_2+M_3\Delta_3+M_4\Delta_4}{1+G_1H_1H_2(G_2-G_3)}$$

$$\therefore \frac{C}{R}=\frac{G_1(G_2-G_3)(1+G_4H_2)}{1+G_1H_1H_2(G_2-G_3)}$$

Example 17:

(i) Find the output for the system of Ex. 1.8 by SFG technique

(ii) If R_1 represents input signal & R_2 represents noise, find signal to noise ratio.

Solution:

(i) **Step 1 :** Mark the nodes on the given block diagram

Step 2 : Write the equation for the nodes: Using Superposition theorem

For R_1 Input alone $y_1 = R_1 - H_1C$; $y_2 = G_1 y_1 - H_3y_3$;

$y_3 = G_2y_2 - H_2C$; $y_4 = y_3$ & $C = G_4y_4$

For R_2 Input alone $y_1 = -H_1C$; $y_4 = y_3 + R_2$. For y_2, y_3 & C the equations will not change.

Step 3 : Draw SFG

Step 4 : Transfer Function

(1) Loop
Loop-gain

(i) $y_1-y_2-y_3-y_4-C-y_1$: $-G_1G_2G_3H_1$

(ii) $y_2-y_3-y_2$: $-G_2H_3$

(iii) $y_3-y_4-C-y_3$: $-G_3H_2$

(2) Two non touching loops : NIL

(3) $\Delta = 1 + [G_1G_2G_3H_1 + G_2H_3 + G_3H_2]$

(4) Forward Path :

(i) For R_1 Input : $R_1 - y_1 - y_2 - y_3 - y_4 - C$

Path gain : $M_1 = G_1G_2G_3$; $\Delta_1 = 1$

(ii) For R_2 Input : $R_2 - y_4 - C$

Path gain : $M_1 = G_3$; $\Delta_1 = 1 + G_2H_3$.

(5) $\dfrac{C_1}{R_1} = \dfrac{G_1G_2G_3}{\Delta}$ &

$$\frac{C_2}{R_2} = \frac{G_3(1+G_2H_3)}{\Delta}$$

$\therefore$ Total output C = output due to R_1 + output due to R_2.

$$C = \frac{(G_1G_2G_3)R_1 + G_3(1+G_2H_3)R_2}{1+[G_2H_3 + G_3H_2 + G_1G_2G_3H_1]}$$

Solution (ii) : Signal $[R_1]$ to Noise $[R_2]$ ratio means we have to find $\dfrac{R_1}{R_2}$.

$$\therefore \frac{R_1}{R_2} = \frac{C}{R_2} \Big/ \frac{C}{R_1} = \frac{G_3[1+G_2H_3]}{\Delta} \Big/ \frac{G_1G_2G_3}{\Delta}$$

$$\therefore \frac{R_1}{R_2} = \frac{1+G_2H_3}{G_1G_2}$$

Example 18:

Find the T.F. from the node equations of Example 1.17

Solution:

Node equations are

$y_1 = R - H_2y_3$ ...(1)

$y_2 = G_1y_1$ ...(2)

$y_3 = H_1C - G_4 R$ ...(3)

$C = (G_2 - G_3)\, y_2$ ...(4)

Put eq. (2) and then (1) into (4)

$\therefore C = (G_2 - G_3)\, G_1 \{R - H_2\, y_3\}$

$= (G_2 - G_3)\, G_1R - G_1\, H_2\, (G_2 - G_3)\, y_3$

Put equation (3) in above expression

$\therefore C = (G_2 - G_3)\, G_1R - G_1H_2\, (G_2 - G_3)\, \{H_1C - G_4R\}$

$= \{(G_2 - G_3)\, G_1 + G_1H_2\, (G_2 - G_3)\, G_4\}\, R - G_1H_2H_1\, (G_2 - G_3)\, C$

$$\therefore \frac{C}{R} = \frac{(G_2 - G_3)(G_1 + G_1G_4H_2)}{1 + G_1H_1H_2(G_2 - G_3)}$$

Procedure : We want the T.F. i.e. $\frac{C}{R}$ hence rearrange the node equation in such a way that the final expression is obtained in terms of C & R.

Example 19:

A control system is represented by the following B.D. Determine the sensitivity of closed loop T.F. with respect to G & H for ω = 1 rad/rec.

Solution:

Sensitivity due to G :

$$S_G^T = \frac{1}{1+GH}$$

$$= \frac{1}{1 + \frac{10 \times 5}{s(s+1)}}$$

$$= \frac{s(s+1)}{s^2 + s + 50}$$

For expressing the T.F. in frequency (ω) domain s is replaced by jω.

$$\therefore S_G^T = \frac{j\omega(j\omega + 1)}{(j\omega)^2 + j\omega + 50} = \frac{j\omega - \omega^2}{50 + j\omega - \omega^2}$$

Put ω = 1

$$\therefore S_G^T = \frac{j-1}{49+j}$$

$$\therefore |S_G^T| = \frac{|j-1|}{|49+j|} = 0.028$$

$$\therefore S_G^T = 0.0288$$

Sensitivity due to H :

$$S_H^T = \frac{-GH}{1+GH} = \frac{10 \times 5 / s(s+1)}{1 + \frac{10 \times 5}{s(s+1)}}$$

$$= \frac{50}{s^2 + s + 50}$$

Put s = jω

$$\therefore\ S_H^T = \frac{50}{(j\omega)^2 + j\omega + 50}$$

$$= \frac{50}{-\omega^2 + j\omega + 50}$$

Put $\omega = 1$;

$$\therefore\ S_H^T = \frac{50}{49 + j}$$

$$\therefore\ \left|S_H^T\right| = 1.02$$

Note : For the same sytem, and for the same frequency

$\left|S_H^T\right| > \left|S_G^T\right|$ i.e. parameter variations in feedback path effect more than the parameter variation in forward path.

Example 20:

For the system shown in the B.D., the normal value of gain k is 25. Evaluate the sensitivity of T.F. $\frac{C}{R}$ due to variation in parameter k for $\omega = 5$ rad/sec.

Solution:

$$G = \frac{k}{s(s+5)} = \frac{25}{(s+5)}$$

H = 1

$$S_k^T = \frac{\partial T}{\partial k} \times \frac{k}{T}$$

$$T = \frac{k}{s^2 + 5s + k}$$

$$\therefore\ \frac{\partial T}{\partial k} = \frac{(s^2 + 5s + k) - k}{(s^2 + 5s + k)^2} = \frac{s(s+5)}{(s^2 + 5s + k)^2}$$

$$S_k^T = \frac{s(s+5)}{(s^2 + 5s + k)^2} \times \frac{k}{k/s^2 + 5s + k}$$

$$= \frac{s(s+5)}{(s^2 + 5s + k)}$$

For k = 25; $$S_k^T = \frac{s(s+5)}{(s^2+5s+25)}$$

Let s = jω

$$\therefore \quad S_k^T = \frac{j\omega(j\omega+5)}{(j\omega)^2+5\times j\omega+25}$$

For ω = 5, $$S_k^T = \frac{25j-25}{25j} = \frac{j-1}{j}$$

$$\therefore \quad \left|S_k^T\right| = 1.41$$

Example 21:

For the following system, find the value of A for 5% sensitivity of T.F. $T(s) = \frac{V_o(s)}{V_i(s)}$ with respect to the variations in μ when μ = 20.

Solution:

$$T(s) = \frac{V_o(s)}{V_i(s)} = \frac{G(s)}{1+G(s).H(s)}$$

$$G(s) = \frac{A}{s+3}\times\frac{0.8\,\mu\, s(s+3)}{s+4}\times\frac{s+4}{s} = 0.8\ \mu\ A$$

$$\therefore \quad T(s) = \frac{0.8\,\mu\,A}{1+0.2\,\mu\,A} \qquad\qquad H(s) = 0.25$$

Now $$S_\mu^T = \frac{\partial T}{\partial \mu}\times\frac{\mu}{T}$$

$$\frac{\partial T}{\partial \mu} = \frac{(1+0.2\,\mu\,A)\,0.8\,A - 0.8\,\mu\,A\,(0.2\,A)}{(1+0.2\,\mu\,A)^2} = \frac{0.8\,A}{(1+0.2\,\mu\,A)^2}$$

$$\therefore \quad S_\mu^T = \frac{0.8\,A}{(1+0.2\,\mu\,A)^2}\times\frac{\mu(1+0.2\,\mu\,A)}{0.8\,\mu\,A} = \frac{1}{1+0.2\,\mu\,A}$$

Given μ = 20 ;

$$S_\mu^T = 5\% = 0.05$$

$$\therefore \quad 0.05 = \frac{1}{1+4\,A} \qquad\qquad \therefore A = 4.75$$

Example 22:

In the speed control system shown in Figure, the generator field time constant is negligible. It is driven at constant speed giving a generated voltage of K_v volts/field amp. The motor is seperately excited so as to have a counter emf of K_b volts per rad/sec. It produces a torque of K_T newton-m/amp. The motor and its load have a combined moment of inertia J $kg - m^2$ and negligible friction. The tachometer has K_t volts per rad/sec and the amplfier gain is K_A amp/volt. Draw the block diagram of this system and determine therefrom the transfer function $\omega(s)/E_i(s)$, where ω is the load speed.

With the system originally at rest, a control voltage $e_i = 100$ volts is suddenly applied. Determine how the load speed will change with time.

Given : $J = 6\ kg - m^2$,

$K_A = 4$ amp/volt,

$K_T = 1.5$ Nm/amp

$K_g = 50$ volts/amp

$K_t = 0.2$ volts per rad/sec

(Hint : $K_b = K_T$ in MKS units)

Solution:

(a) Construction of Block Diagram

This speed control problem involves following components :

(i) D.C. Motor in armature control mode.

(ii) Separately excited D.C. Generator for controlling the motor speed.

(iii) Load driven by motor.

(iv) Tachometer as a speed sensor hence it will be in feed back path.

(v) Potentiometer as error detector.

(vi) Amplifier.

We already know the equivalent B.D. & T.F. for all these components. (Refer to previous sections). Hence we can replace these components by their B.D. and join them according to their connection in the system.

Here

Input = reference voltage (e_i) to the pot

Output = speed (ω)

Feedback = output of tachometer

(b) Transfer Function

Step 1 : Solve the internal loop first

$$G(s) = \frac{1}{R_a} \times K_T \times \frac{1}{J_s} = \frac{K_T}{R_a \cdot J \cdot s}$$

$$H(s) = K_b$$

$$T.F. = \frac{G(s)}{1 + G(s).H(s)}$$

$$\therefore \; T.F. = \frac{\omega(s)}{V_a(s)} = \frac{K_T}{R_a J_s + K_T K_b}$$

Step 2 : Solve the external feed backloop

$$G(s) = \frac{K_T}{R_a J_s + K_T \cdot K_b} \times K_g \times K_A$$

$$H(s) = K_t.$$

$$\therefore \; T.F. = \frac{\omega(s)}{E_i(s)} = \frac{K_T \cdot K_g \cdot K_A}{R_a \cdot J \cdot s + [K_T K_b + K_T \cdot k_g \cdot K_A \cdot K_t]}$$

Now substitute the values

$$\therefore \; \frac{\omega(s)}{E_i(s)} = \frac{1.5 \times 50 \times 4}{(1 \times 6 \times s) + [1.5 \times 1.5 + 1.5 \times 50 \times 4 \times 0.2]}$$

$$\frac{300}{6s + 62.25} = \frac{300}{6\left[s + \frac{62.25}{6}\right]}$$

$$\therefore \; \frac{\omega(s)}{E_i(s)} = \frac{50}{s + 10.375}$$

(c) Here we have to find $\omega(t)$ when

$$E_i(s) = \frac{100}{s} \text{ Volts [It is a step input]}$$

$$\therefore \; \omega(s) = \left\{\frac{50}{s + 10.375} \times \frac{100}{s}\right\}$$

$$\therefore \; \omega(t) = \ell^{-1}\left[\frac{5000}{s(s + 10.375)}\right]$$

$$\therefore \; \omega(t) = 482\left[1 - e^{-10.375t}\right] \text{rad/sec}$$

Example 23:

Consider the positional servomechanism shown in figure below. Assume that the input to the system is the reference shaft position θ_R and the system output is the load shaft position θ_L. Draw the block diagram of the system indicating the transfer function of each block. Simplify the block diagram to obtain $\theta_L(s)/\theta_R(s)$ for the closed-loop system and also when the loop is open (in opening the loop, the lead from the output potentiometer driven by θ_C is disconnected and grounded). The parameters of the system are given below:

Sensitivity of error detector K_P = 10 volt/rad

Gain of d.c. amplifier K_A = 50 volts/volt

Moter field resistance R_f = 100 ohms

Moter field inductance L_f = 20 henrys

Motor torque constant K_T = 10 newton-m/amp

Moment of inertia of load J_L=250 kg–m^2

Cofficient of viscous friction of load f_L = 2500 newton-m per rad sec

Motor to load gear ratio $(\dot{\theta}_1/\dot{\theta}_M) = 1/50$

Load to potentiometer gear ratio $(\dot{\theta}_C/\dot{\theta}_L) = 1$

Motor inertia and friction are negligible

Solution:

(a) In this example D.C. Motor in field control mode is used for position control system. Here the position of a load is controlled by D.C. Motor with the help of gear. It is sensed by potentiometer again with the help of gear. Output of the potentiometer is amplified & given to the field circuit. Hence the components are :

(i) D.C. Motor in field-controlled mode

(ii) Amplifier

(iii) Potentiometer as sensor device

(iv) Two Gear Mechanism

The B.D. will be as under :

Transfer Function

(i) For open-Loop mode (Switch is open)

$$G(s) = K_P \times K_A \times \frac{1}{R_f + sL_f} \times K_T \times \frac{1}{J_m s + f_m} \times \frac{1}{s} \times \frac{1}{50}$$

J_m = Moment fo inertia of load referred to motor side

$$= \left[\frac{N_1}{N_2}\right]^2 J_L$$

f_m = Viscous friction coefficient of load referred to motor side

$$= \left[\frac{N_1}{N_2}\right]^2 f_L$$

$$\frac{N_1}{N_2} = \text{Load to motor gear ratio} = \frac{\dot{\theta}_L}{\dot{\theta}_m} = \frac{1}{50}$$

$$\therefore J_m = \left(\frac{1}{50}\right)^2 \times 250 = 0.1$$

$$f_m = \left(\frac{1}{50}\right)^2 \times 2500 = 1$$

$$\therefore \text{T.F.} = \frac{\theta_L}{\theta_R} = G(s)$$

$$= 10 \times 50 \times \frac{1}{100+20s} \times 10 \times \frac{1}{1+0.1s} \times \frac{1}{s} \times \frac{1}{50}$$

$$= \frac{100}{s(0.1s+1)(20s+100)} = \frac{50}{s(s+10)(s+5)}$$

$$\text{T.F.} = G(s) = \frac{\theta_L}{\theta_R} = \frac{50}{s(s+10)(s+5)}$$

(ii) **Closed Loop T.F. (Switch is closed):**

$$\text{T.F.} = T(s) = \frac{\theta_L}{\theta_R} = \frac{G(s)}{1+G(s).H(s)} ;H(s) = 1$$

$$\therefore T(s) = \frac{50/s(s+5)(s+10)}{1+\dfrac{50\times 1}{s(s+5)(s+10)}} = \frac{50}{s^3+15s^2+50s+50}$$

$$\therefore T(s) = \frac{\theta_L}{\theta_R} = \frac{50}{s^3+15s^2+50s+50}$$

Example 24:

A two phase servo-motor has following specifications : Starting torque (T_{st}) = 0.166 N-m; Supply voltage [v_c] = 115 volts, No load speed (N_o) = 2904 rpm; Moment of Inertia (J) = 10^{-5} kg – m^2; Friction is neglected.

(a) Find the transfer function.

(b) Draw the Torque v/s speed characteristics for the control winding voltage of 57.5 volts.

(c) How the speed will vary when the control winding voltage undergoes a unit step charge ?

(d) What will be the steady state speed?

Solution:

(a) For two phase servo - motor;

$$T.F. = \frac{\theta_m(s)}{V_c(s)} = \frac{K_m}{s[1+s\tau_m]}$$

Where $K_m \frac{K}{f_o+f}$ & $\tau_m = \frac{J}{f_0+f}$;

$$K = \frac{T_0}{V_c}; f = \frac{T_0}{\omega_0}$$

K_m = Motor gain;

τ_m = Motor time constant.

T_0 = Starting torque,

ω_0 = No load speed = $2\pi N_0$; (N_o in rps)

v_c = Rated control winding voltage

$$\therefore K = \frac{0.166}{115} = 1.44 \times 10^{-3};$$

$$f = \frac{0.166}{2\pi \times \frac{2904}{60}} = 5.46 \times 10^{-4}$$

$$K_m = \frac{1.44 \times 10^{-3}}{5.46 \times 10^{-4}} = 2.66$$

$$\tau_m = \frac{10^{-5}}{5.46 \times 10^{-4}} = 0.0183$$

$$\therefore T.F. = \frac{\theta_m(s)}{V_c(s)} = \frac{2.66}{s(1+0.0183\,s)}$$

(b) Draw the torque v/s speed characteristics for the given data i.e. for rated controlling voltage.

Solution:

The characteristic is linear. The characteristic for other valves of control winding voltages, will be parallel to this characteristic.

As $V_c = 57.5$ i.e. $\frac{V_{c(rated)}}{2}$

$$\therefore \; T_{st} = \frac{0.166}{2} = 0.083$$

$$\& \; N_0 = \frac{2904}{2} = 1452 \text{ rpm}$$

(c) We have, $\frac{\theta_m(s)}{V_c(s)} = \frac{2.66}{s(1+0.0183s)}$

$$v_c(s) = \text{Unit Step} = 1/s$$

$$\therefore \; \theta(s) = \frac{2.66}{s(1+0.0183s)} \times \frac{1}{s}$$

$$\text{i.e.} \; s.\theta(s) = \frac{2.66}{s(1+0.0183s)}$$

$$s.\,\theta(s) = \dot{\theta}(s) = \omega(s) = \text{Speed}$$

$$\therefore \; \omega(s) = \frac{2.66}{s(1+0.0183s)}$$

$$\omega(t) = £^{-1}[\omega(s)] = 145.35\,[1 - e^{-0.0183\,t}]$$

(d) Steady state speed ω_{ss} is obtained by

$$\lim_{t \to \infty} \omega(t)$$

$$\therefore \; \omega_{ss} = \lim_{t \to \infty} [145.35\,(1 - e^{-0.0183\,t}]$$

$$\therefore \; \omega_{ss} = 145.35 \text{ rad/sec}$$

Example 25(a):

The schematic diagram of a servo system is shown in figure below. The two-phase servomoter develops a torque in accordance with the equation

$$T_M = -K_1\,\dot{\theta}_M + K_2\,v_c$$

Where, $K_1 = 0.5 \times 10^{-3}$ newton-m per rad. sec

$K_2 = 2 \times 10^{-3}$ newton-m/volt

Draw the block diagram of the system indicating the trasfer function of each block. Calculate the value of amplifier gain K_A such that system has a steady-state error $(\theta_R - \theta_C)$ of 0.1 deg for an input velocity of 1 rad/sec. What will be the control field voltage under this condition ? Given :

Load inertia	J_L	=	2.5 kg – m²
Coefficient of load friction	f_L	=	250 newton-m per rad/sec.
Motor to load gear ration	$(\dot{\theta}_L/\dot{\theta}_M)$	=	1/50
Motor to synchro gear ratio	$(\dot{\theta}_C/\dot{\theta}_L)$	=	1/1
Sensitivity of synchro	K_s	=	100 volts/rad

The motor inertia and friction are negligible.

Solution:

(A) In this physical system an a.c. servo-motor is driving a load through gear. The position of load is sensed by synchro through gear. The position is controlled by synchro generator. The proportionated output voltage of synchros is given to the control winding through amplifier. Therefore components used are :

(i) A.C. Servo - motor, (ii) Amplifier, (iii) Synchros, (iv) Gears

Note : Block Diagram of Servo motor is obtained according to its torque equation

$$T_m = -K_1\,\dot{\theta}_M + K_2\,v_c$$

$$J_{eq} = \left(\frac{N_1}{N_2}\right)^2 J_L\,;\ \frac{N_1}{N_2} = \frac{\dot{\theta}_L}{\dot{\theta}_m} = \frac{1}{50}\,;\quad J_{eq} = \left(\frac{1}{50}\right)^2 \times 2.5 = 10^{-3}$$

$$f_{eq} = \left(\frac{N_1}{N_2}\right)^2 f_L = \left(\frac{1}{50}\right)^2 \times 250 = 0.1$$

The complete B.D. will be as under

Transfer Function :

Step 1 : Solve the internal feedback loop

$$\frac{\dot{\theta}_m}{K_2\,v_c} = \frac{1/J_{eq}\,.s + f_{eq}}{1 + K_1\,.\dfrac{1}{J_{eq}\,s + f_{eq}}}\qquad \frac{\dot{\theta}_m}{v_c} = \frac{K_2}{J_{eq}\,.s + (f_{eq} + K_1')}$$

$$= \frac{2\times10^{-3}}{10^{-3}\,s + (0.1 + 0.5\times10^{-3})} = \frac{2}{s + 100.5}$$

Step 2 : Solve the outer feedback loops

$$G(s) = K_s \times K_A \times \frac{\dot{\theta}_m}{v_c} \times \frac{1}{s} \times \frac{1}{50} \times \frac{1}{50} \text{ \& upto}$$

$$\theta_C = K_s \times K_a \times \frac{\dot{\theta}_m}{v_c} \times \frac{1}{s} \times 1$$

H(s) = 1

$$\therefore \frac{\theta_c}{\theta_R} = \frac{K_s K_A \times \frac{2}{s+100.5} \times \frac{1}{s} \times 1}{1 \times K_s . K_A \times \frac{2}{s+100.5} \times \frac{1}{s} \times 1}$$

$$= \frac{2 K_s . K_A}{s(5+100.5)+2 K_s . K_A}$$

$$\therefore \frac{\theta_c}{\theta_R} = \frac{200 K_A}{s(5+100.5)+200 K_A} \quad ...(1)$$

(B) From the B.D.;

$$\frac{\theta_c}{\theta_R - \theta_c} = K_s \times K_A \times \frac{\dot{\theta}_m}{v_c} \times \frac{1}{s} \times 1$$

$$= 100 \times K_A \times \frac{2}{s+100.5} \times \frac{1}{s} \times 1$$

$$= \frac{200 K_A}{s(s+100.5)} \quad ...(2)$$

$$\frac{(1)}{(2)} = \frac{\theta_c/\theta_R}{\theta_c/(\theta_R - \theta_C)}$$

$$= \frac{200 K_A / s^2 + 100.5 s + 200 K_A}{200 K_A / (s+100.5) s}$$

$$\therefore \frac{\theta_R - \theta_C}{\theta_R} = \frac{s(s+100.5)}{s^2 (100.5 s + 4 K_A)}$$

Given ;

$$\theta_R = 1 \times \frac{1}{s} \text{ \& } \theta_R - \theta_C = 0.1 \text{ degree}$$

$$= 0.1 \times \frac{\pi}{180} \text{ rad}$$

This is when s → 0 i.e. Steady state condition

$$\therefore \frac{0.1\times\pi}{180} = \frac{s(s+100.5)}{s^2+100.5s+200K_A}\times\frac{1}{s}\Big|_{s\to 0}$$

$$\therefore 0.1\times\frac{\pi}{180} = \frac{100.5}{200\ K_A}$$

$$\therefore K_A = \frac{100.5\times 180}{200\times\pi\times 0.1} = 287.91$$

$\therefore K_A = 287.91$

(C) With this value of K_A

$$\frac{\theta_c}{\theta_R} = \frac{200\times 287.91}{s(s+100.5)+200\times 287.91}$$

$$\therefore \theta_c = \frac{200\times 287.91}{s(s+100.5)+(200\times 287.91)}\times\frac{1}{s} \qquad \left[\text{As } \theta_R = \frac{1}{s}\right]$$

$$\therefore s.\theta_c = \dot{\theta}_c = \frac{200\times 287.91}{s(s+100.5)+(200\times 287.91)}$$

Also $\dot{\theta}_m = \dot{\theta}_c$

$$\frac{\theta_m}{v_c} = \frac{2}{s+100.5}$$

$$\therefore v_c = \left(\frac{s+100.5}{2}\right)\times\dot{\theta}_m = \frac{s+100.5}{2}\times\dot{\theta}_c$$

∴ v_c for steady state condition [s → 0] will be

$$v_c = \frac{0+100.5}{2}\times\frac{200\times 287.91}{0(0+100.5)+(200\times 287.9)} = 50.25$$

∴ v_c = 50.25 Volts.

Example 25(b):

In the position control system shown in figure below, the sensitivity of the synchro error detector is 1 volt/deg. The transfer function of the motor is found to be

$$\frac{\theta_M(s)}{v_c(s)} = \frac{K_m}{s(s\tau_m+1)}$$

Where K_m=15rad/sec/volt; τ_m=0.15 sec

The gear ratios are given as $\dot{\theta}_C / \dot{\theta}_L = 1 \quad \dot{\theta}_L / \dot{\theta}_M = 1/50$

(a) Draw the Block Diagram

(b) Find the Transfer Function.

Solution:

(a) In this sytem an a.c. servo motor is positioning a load. A gear is used to reduce the speed. The positon is sensed by the synchro. The equivalent output voltage of synchro is amplified & it is fed to the control winding of the servo-motor. Hence its speed is controlled.

∴ The components are :

(i) A.C. Servo motor (ii) Synchros (iii) Gears (iv) Amplifier

Therefore the B.D. will be as under

Note : Given $\dfrac{\theta_m}{v_c} = \dfrac{K_m}{s(1+\tau_m s)}$ for motor. This can also be written as

$$\frac{s\theta_m}{v_c} = \frac{\dot{\theta}_m}{v_c} = \frac{K_m}{(1+\tau_m s)}$$

(b) For the forward path;

$$G(s) = K_s \times K_A \times \frac{K_m}{1+\tau_m s} \times \frac{1}{50} \times \frac{1}{s}$$

$$\therefore \quad G(s) = \frac{K_s . K_A . K_m}{50\, s(1+\tau_m s)} = \frac{\left(1 \times \dfrac{180}{\pi}\right) \times 10 \times 15}{50\, s(1+0.15\, s)}$$

$$= \frac{171.89}{s(1+0.15\, s)}$$

[K_s = 1 volt/deg. ∴ In radians; $K_s = \left(1 \times \dfrac{180}{\pi}\right)$ volts/rad]

$$\therefore \quad G(s) = \frac{171.89}{s(1+0.15\, s)}$$

$$H(s) = 1$$

$$\therefore \quad T(s) = \frac{G(s)}{1+G(s).H(s)} = \frac{\theta_L}{\theta_R} = \frac{171.89/ s(1+1.15\, s)}{1+\dfrac{171.89}{s(1+0.15\, s)}}$$

$$\therefore \frac{\theta_L}{\theta_R} = \frac{171.89}{0.15s^2 + s + 171.89}$$

$$\therefore \frac{\theta_L}{\theta_R} = \frac{1146}{s^2 + 6.67s + 1146}$$

Example 26:

The schematic diagram of a servo-mechanism is given in figure below. The system constans are as follows :

Synchro sensitivity K_s = 1 volt/rad

Amplifier gain K_A = 20 volt/volt

Motor torque constant $K_T = 10^{-5}$ newton-m/volt

Load inertia $J_l = 1.5 \times 10^{-5}$ kg – m^2

Viscous friction $f_L = 1 \times 10^{-5}$ newton-m/rad/sec

Tachometer constant K_t = 0.2 volt/rad/sec

(a) Draw the Block Diagram

(b) Find the Transfer Function.

Solution:

(a) A.C. Servo - meter is driving a load. Speed is sensed by a tachogeneator. Synchro is used as an error - detector. The equivalent voltage is fed to the control winding of the motor through amplifier. At the Input of Amp. Output of synchro & tachometer are connected. Hence its speed is controlled. The components are

(i) A.C. Servo - motor; (ii) Synchro; (iii) Tachometer

(iv) Gear; (v) Load

For motor

J = Load inertia referred to the motor side = J_L.

f = Viscous friction referred to the motor side = f_L

Hence the Block diagram will be as under

(b) Transfer Function

Step 1 : Simplify the internal f/b loop first :

$$\frac{\dot{\theta}_m}{e_s} = \frac{K_A \times K_T / J_s + f}{1 + \dfrac{K_A \times K_T \times K_t}{J_s + f}} = \frac{K_A . K_T}{J_s + [K_A . K_T . K_t + f]}$$

Step 2 : Solve the outer feedback loop.

$$\frac{\theta_m}{\theta_R} = \frac{\{K_s \times K_A . K_T \times 1/s\}/\ J_s + (K_A . K_T . K_t + f)\}}{1 + \dfrac{K_s \times K_A . K_T \times 1/s}{J_s + (K_A . K_T . K_t + f)} \times 1}$$

$$\therefore \qquad T.F. = \frac{\theta_m}{\theta_R} = \frac{K_s . K_A . K_T}{Js^2 + (K_A . K_T . K_t + f)s + K_s . K_A . K_T}$$

$$\frac{\theta_m}{\theta_R} = \frac{20}{1.5\,s^2 + 5\,s + 20}$$

$$\therefore \qquad \frac{\theta_m}{\theta_R} = \frac{13.33}{s^2 + 3.33\,s + 13.33}$$

Example 27(a):

A simple voltage regulator is shown in figure below. A potentiometer is used at the output terminals of the generator to give a feedback voltage K v_0 where K is constant ($K \leq 1$). The potentiometer resistance is high enough that it may be assumed to draw negligible current. The amplifier has a gain of 20 volts/volt. The generator gain is 50 volts/field amp. Reference voltage v_r = 50 volts.

(a) Draw a block diagram of the system when the generator is supplying a load current indicate the transfer function of each block.

(b) The system is operated closed-loop (s closed). Determine the value of K in order to give a steady no-load generator terminal voltage of 250 volts. What is the change in terminal voltage caused by a steady load current of 30 amps ? What reference voltage would be required to restore the generator terminal voltage of 250 V ?

(c) The system is operated open-loop ('s') open, determine the reference voltage needed to obtain a steady no-load voltage of 250 V. What would be the change in terminal voltage for a load of 30 amps ?

(d) Compare the changes in terminal voltage in parts (b) and (c) and comment upon the effect of feedback in countering the changes in terminal voltage caused by load current.

Solution:

(a) This is a voltage regulator system. Generator terminal voltage is controlled by the field current. Field circuit is separately excited. Generator terminal voltage is sensed by the potentiometer & it is fed back. Field circuit is energised through amplifier. Hence the componets are :

(i) D.C. Generator (Separately excited)

(ii) Potentiometer; (iii) Amplifier.

Note : For D.C. Generator, Input is output of the amplifier. It is given to the field circuit. Generated emf say E_g is proportional to I_f. For generator, terminal voltage V_0 is given as $V_0 = E_g - I_a R_a$ where I_a = Armature current & R_a = Armature resistance.

Hence for generator B.D. will be as shown.

Transfer Functions

$$V_0 = E_g - I_a R_a \quad ...(1)$$

$$E_g = K_g . I_f \quad ...(2)$$

$$I_f = \frac{(V_r - K V_0) . K_A}{R_f + s L_f} \quad ...(3)$$

(b) (i) System is operating at No - load

V_0 = 250 Volts;

Switch 'S' is closed.

$\therefore I_a = 0$

From (1), (2) and (3) we get

$$V_0 = E_g - 0 \times R_a = K_g \times I_f$$

$$= K_g \left[\frac{V_r - K V_0}{R_f + s L_f}\right] K_A \quad ...(4)$$

For steady state condition, s = 0

$$\therefore V_0 = K_g \left[\frac{V_r - K V_0}{R_f + s L_f}\right] K_A$$

$$250 = 50 \left[\frac{50 - K \times 250}{100}\right] 20 \quad \therefore K = 0.1$$

(ii) Now Load Current = $I_L = I_a$ = 30 Amp

$$\therefore V_0 = E_g - I_a . R_a.$$

$$= K_g \left[\frac{V_r - K V_0}{R_f}\right] K_A - I_s . R_a \quad ...(5)$$

$$= 50\left[\frac{50 - 0.1\ V_0}{100}\right] 20 - 30 \times 1$$

$\therefore\ V_0 = 235$ Volts

$\therefore$ **Drop in terminal voltage = 15 volts**

(iii) $V_0 = 250$ Volts $V_r = ?$ Using eq. (5)

$$250 = 50\left[\frac{V_r - 0.1 \times 250}{100}\right] 20 - 30 \times 1$$

$\therefore\ V_r = 53$ Volts

(c) Switch 'S' is open. Hence system is operating in open - loop mode.

(i) $I_a = 0;\ V_0 = 250$ Volts;

For steady state condition s = 0 ;

From the Block Diagram if 'S' is open

$V_0 = E_g - I_a \cdot R_a.\qquad E_g = K_g\ I_f$

$$\therefore\ I_f = \frac{V_r \cdot K_A}{R_f + sL_f}$$

$$\therefore\ V_0 = K_g \frac{V_r \cdot K_A}{R_f} - I_a R_a \qquad ...(6)$$

$$\therefore\ 250 = 50 \times \frac{V_r \times 20}{100} \qquad V_r = 25 \text{ volts}$$

(ii) $I_a = 30$ Amp — Using equation (6)

$$V_0 = \frac{50 \times 25 \times 20}{100} - 30 \times 1 = 220 \text{ Volts}$$

$\therefore$ **Change in terminal voltage = 30 Volts**

(d) From the result of part (b) and (c) it is clear that when system is operating in closed loop mode the drop in terminal voltage due to the same load current is less.

Example 27(b):

For the speed control system shown in figure below assume that

(i) The reference and feedback tachometers are identical.

(ii) Generator field time constant is negligible and its generated voltage is K^o volts/amp.

(a) Draw the block diagram and obtain the T.F. when the system is operating in open loop & in closed loop mode.

(b) Find the time variation of output speed (ω_0) for a sudden reference input of 10 rad/sec. Find also the steady-state output speed.

(c) If the feedback loop is opened and gain K_A adjusted to give the same steady-state speed as in the case of the closed-loop, determine how the output speed varies with time and compare the speed of response in two cases.

(Hint : The generator gain constant K_g changes in direct proportion to generator speed, i.e. $K_g = K_g' \omega_g$, where K_g is a constant)

The system constants are given below:

Moment of inertia of motor and load J = 5 Kg – m^2

Motor back emf constant K_i = 5 volts per rad/sec

Total armature resistance of motor and generator R_a = 1 ohm

Generator gain constant K_g = 50 volts/amp

Amplifier gain K_A = 5 amps/volt

Tachometer constant K_i = 0.5 volts per rad/sec

Solution:

(a) In this speed control system motor is driving a load. Motor is operating in armature controlled mode. Armature current is controlled by the generator voltage which in turn is controlled by the field current. One tachometer is used as ref. & other as feed back.

The output of these two tachometers is compared & after amplification given to field circuit of the generator. Hence the components are

(i) D.C. Motor

(ii) D.C. Generator

(iii) Feed back Tachometer

(iv) Reference Tachometer

(v) Amplifier

(vi) Load

The input is reference speed ω_r & the output is load speed ω_0.

Note : If K_T is given & K_b is not given; assume $K_b = K_T$ & vice versa.

Transfer Function when switch is open

For Internal Loop :

$$G(s) = \frac{1}{R_a} \times K_T \times \frac{1}{J.s} \qquad \& \ H(s) = K_b$$

$$\therefore \frac{\omega_0}{E_g} = \frac{\frac{1}{R_a} \times K_T \times \frac{1}{J.s}}{1 + \frac{1}{R_a} \times K_T \times \frac{1}{J.s} \times K_b}$$

$$\therefore \frac{\omega_0}{E_g} = \frac{K_T}{R_a J.s + K_T.K_b}$$

For the complete forward path

$$G(s) = \frac{K_t.K_A.K_g.K_T}{R_a.J.s + K_T.K_b}$$

∴ Open Loop

$$T.F. = \frac{\omega_0}{\omega_r} = \frac{K_t.K_A.K_g.K_T}{R_a.J.s + K_T.K_b} \qquad ...(1)$$

$$\therefore \frac{\omega_0}{\omega_r} = \frac{0.5 \times 50 \times 5 \times 5}{(1 \times 5)s + (5 \times 5)} = \frac{625}{5s + 25}$$

$$\therefore \frac{\omega_0}{\omega_r} = \frac{125}{s+5} \qquad ...(2)$$

When Switch is closed : H(s) = 1

$$\therefore \frac{\omega_0}{\omega_r} = \frac{\frac{K_A.K_t.K_g.K_T}{R_a.Js + K_T.K_b}}{1 + \frac{K_A.K_t.K_g.K_T}{R_a.Js + K_T.K_b}}$$

$$\frac{\omega_0}{\omega_r} = \frac{K_A.K_t.K_g.K_T}{R_a.Js + K_A.K_T.K_g.K_T + K_T.K_b} \qquad ...(3)$$

$$\therefore \frac{\omega_0}{\omega_r} = \frac{5 \times 0.5 \times 50 \times 5}{(1 \times 5)s + (5 \times 0.5 \times 50 \times 5 + 5 \times 5)} = \frac{625}{5s + 650}$$

$$\therefore \frac{\omega_0}{\omega_r} = \frac{125}{s+130} \qquad ...(4)$$

(b) Given $\omega_r = \frac{10}{s}$ rad/sec

Put in equation (4)

$$\therefore \quad \omega_o = \frac{125}{s+130} \times \frac{10}{s} = \frac{1250}{s(s+130)}$$

$$\therefore\ \omega_o(t) = L^{-1}\left[\frac{1250}{s(s+130)}\right]$$

$$\therefore\ \omega_0(t) = 9.6\ [1 - e^{-130\,t}]$$

Steady state speed;

$$(\omega_o)_{ss} = \lim_{t\to\infty} \omega_o(t)\quad \therefore\ (\omega_0)_{ss} = 9.6 \text{ rad/sec}$$

(c) Switch is open & K_A is adjusted to give $(\omega_0)_{ss}$ = 9.6 rad/sec

Equation (1) is

$$\frac{\omega_o}{\omega_r} = \frac{K_t . K_A . K_g . K_T}{J . R_a . s + K_T . K_b}$$

$$= \frac{0.5\times 50\times 5\times K_A}{(1\times 5)s+(5\times 5)} = \frac{25\,KA}{s+5} \quad ...(5)$$

$$\omega_r = \frac{10}{s}\ \therefore\ \omega_o = \frac{25\,K_A}{s+5}\times\frac{10}{s}$$

$$(\omega_o)_{ss} = \lim_{s\to 0} s.[\omega_o] = \lim_{s\to 0} s\times\frac{250\,K_A}{s(s+5)} = 50\ K_A$$

$(\omega_o)_{ss}$ = 9.6 rad/sec (Given)

$\therefore$ 50 K_A = 9.6

$\therefore$ K_a = 0.1923

To find variation of speed put this value of K_A in equation (5) for

$$\omega_r = \frac{10}{s}$$

Take Inverse laplace transform

$$\therefore\ \omega_o = \frac{25\times 0.1923}{s+5}\times\frac{10}{s} = \frac{48}{s(s+5)}$$

$$\therefore\ \omega_o(t) = L^{-1}\left[\frac{48}{s(s+5)}\right]$$

$$\omega_o(t) = 9.6\left[1-e^{-5t}\right]$$

If we compare the equation of response in open loop & closed loop mode we find that speed of response in closed loop mode is much higher [Time constant = $\frac{1}{130}$] than that in open loop mode [Time constant = $\frac{1}{5}$] Hence due to feedback speed of response improves.

Example 28:

Find out the T.F. of the circuit given below.

Solution:

Step 1 : Draw the Laplace - Transform of the circuit.

Step 2 : Apply KVL

$$E_i(s) = \left[R + sL + \frac{1}{sC}\right] I(s) \qquad ...(1)$$

$$E_o(s) = \frac{1}{sC} I(s)$$

$$\therefore \quad T.F. = \frac{E_o(s)}{E_i(s)} = \frac{1/sC}{R + sL + 1/sC}$$

$$= \frac{1}{s^2 LC + sRC + 1}$$

$$T.F. = \frac{1/LC}{s^2 + \frac{R}{L}s + \frac{1}{LC}}$$

Example 29:

Find the T.F. of the Lead Network.

Solution:

Step 1 : Draw the L.T. of the circuit

Step 2 : Apply KVL

$$E_i(s) = \left[\left\{R_1 || \frac{1}{sC}\right\} + R_2\right] I(s)$$

$$= \left[\frac{R_1 \cdot \frac{1}{sC}}{R_1 + \frac{1}{sC}} + R_2\right] I(s) = \left[\frac{R_1}{1 + R_1 sC} + R_2\right] I(s)$$

$$= \left\{\frac{R_1 + R_2 + R_1 . R_2 sC}{1 + R_1 sC}\right\} I(s)$$

$E_0(s) = R_2\ I(s)$

$$\therefore\ T.F. = \frac{E_o(s)}{E_i(s)} = \frac{R_2(1+R_1 sC)}{R_1+R_2+R_1 . R_2 sC}$$

$$\therefore\ T.F. = \frac{s+\frac{1}{R_1 C}}{s+\left[\frac{1}{R_1 C}+\frac{1}{R_2 C}\right]}$$

Example 30:

Find the T.F. of Lag Network given below

Solution:

Step 1: Draw the L.T. of the Network

Step 2 : Apply KVL

$$E_i(s) = \left[R_1+R_2+\frac{1}{sC}\right] I(s)$$

$$= \frac{1+(R_1+R_2)sC}{sC} I(s)$$

$$E_o(s) = \left(R_2+\frac{1}{sC}\right) I(s) = \frac{1+R_2 sC}{sC} I(s)$$

$$\therefore\ T.F. = \frac{E_o(s)}{E_i(s)} = \frac{1+R_2 sC}{1+(R_1+R_2)sC}$$

$$= \frac{R_2 C\left[s+\frac{1}{R_2 C}\right]}{(R_1+R_2)C\left\{s+\frac{1}{(R_1+R_2)C}\right\}}$$

$$\therefore\ T.F. = \left\{\frac{R_2}{R_1+R_2}\right\}\left[\frac{s+\frac{1}{R_2 C}}{s+\frac{1}{(R_1+R_2)C}}\right]$$

Example 31:

A servo system is represented by the signal flow graph shown in fig. The variable T is the torque and E is the error. Determine:

(a) The overall T.F. if $K_1=1$, $K_2 = 5$ and $K_3 = 5$

(b) The sensitivity of the system to changes in K_1 for $\omega = 0$.

Solution:

(a) Find T.F. $T_e = \dfrac{\theta_o}{\theta_i}$ by Mason's Gain Formula

1. **Loop**

Loop-Gain

(i) $E_c - E_m - T - E_c$

$$\frac{-K_1 K_2}{1+s}$$

(ii) $E_m - T - \theta_0 - E_m$

$$\frac{-5}{s^2 (1+s)}$$

(iii) $E - E_c - E_m - T - \theta_0 - E$ $\qquad \dfrac{-K_1 K_3}{s^2 (s+1)}$

2. Two Non-touching loops : NIL

3. $$\Delta = 1+\left[\frac{K_1K_2}{s+1}+\frac{5}{s^2 (s+1)}+\frac{K_1K_3}{s^2(s+1)}\right]$$

$$= 1+\left[\frac{K_1K_2 s^2 +5+K_1 K_3}{s^2 (s+1)}\right]$$

4. **Forward**
Path Gain M_x $\quad \Delta_x$.

(i) $\theta_i - E - E_c - E_m - T - \theta_0$; $\qquad \dfrac{K_1 . K_3}{s^2 (s+1)}$;

5. $$T = \frac{\theta_0}{\theta_i} = \frac{M_1 . \Delta_1}{\Delta} = \frac{K_1 . K_3/s^2 (s+1)}{1+\dfrac{K_1K_2 s^2 +K_1K_3 +5}{s^2 (s+1)}}$$

$$T = \frac{\theta_0}{\theta_i} = \frac{K_1 K_3}{s^3 + (1 + K_1 K_2)s^2 + (K_1 K_3 + 5)}$$

Given $K_1 = 1$, $K_2 = 5$ & $K_3 = 5$

$$\therefore \quad T = \frac{\theta_0}{\theta_i} = \frac{5}{s^3 + 6s^2 + 10}$$

Solution : (b)

$$S_{K_1}^T = \frac{\partial T}{\partial K_1} \times \frac{K_1}{T}$$

$$T = \frac{5K_1}{s^3 + (1 + 5K_1)s^2 + (5 + 5K_1)}$$

$$\frac{\partial T}{\partial K_1} = \frac{5\{s^3 + (1 + 5K_1)s^2 + (5 + 5K_1)\} - 5K_1\{5s^2 + 5\}}{\{s^2 + (1 + 5K_1)s^2 + (5 + 5K_1)\}^2}$$

$$S_{K_1}^T = \frac{5s^3 + 5s^2 + 25}{\{s^3 + (1 + 5K_1)s^2 + (5 + 5K_1)\}^2} \times \frac{K_1\{s^3 + (1 + 5K_1)s^2 + (5 + 5K)}{5K_1}$$

Put $s = j\omega$, $\omega = 0$; $K_1 = 1$

$$\therefore \quad S_{K_1}^T = \frac{5}{5 + 5 \times 1} = 0.5$$

$$\therefore \quad S_{K_1}^T = 0.5$$

Example 32:

Find the T.F. of the following circuit

Solution:

Step 1 : Draw the L.T. of the circuit

Step 2 : Apply KVL to first loop

$$E_i(s) = R\, I_1(s) + \frac{1}{sC}\,[I_1(s) - I_2(s)] \qquad \text{...(i)}$$

Step 3 : Apply KVL to second loop

$$E_o(s) = -R\, I_2(s) + \frac{1}{sC}\,[I_1(s) - I_2(s)] \qquad \text{...(ii)}$$

Also $E_o(s) = \frac{1}{sC} I_2(s)$

$\therefore \quad I_2(s) = sC\ E_0(s)$ ...(iii)

Put into (i)

$$\therefore \quad E_i(s) = \left[R + \frac{1}{sC}\right] I_1(s) - E_0(s)$$

$$\therefore \quad I_1(s) = \frac{E_i(s) + E_0(s)}{R + 1/sC} \quad \text{...(iv)}$$

Put (iii) & (iv) into (ii)

$$\therefore \quad E_0(s) = -R.sCE_0(s) - E_0(s) + \frac{1}{sC}\left\{\frac{E_i(s) + E_0(s)}{R + \frac{1}{sC}}\right\}$$

$$E_0(s)\left[2 + RsC - \frac{1}{1 + RsC}\right] = \frac{E_i(s)}{1 + RsC}$$

$$\therefore \quad T.F. = \frac{E_0(s)}{E_i(s)} = \frac{1}{2R^2 s^2 C^2 + 3RsC + 1}$$

$$\therefore \quad T.F. = \frac{E_0(s)}{E_i(s)} = \frac{1/R^2C^2}{s^2 + \frac{3}{RC}s + \frac{1}{R^2C^2}}$$

Example 33:

A control system is represented by the following block diagram. Find the T.F., $\frac{C(s)}{R(s)}$.

Solution:

Step 1 : Shift the take - off point of feedback element H_2 ahead of block G_3. Also represent the same feedback element separately for both the sum devices.

$\therefore$ B.D. reduces as

Step 2 : Now consider the internal loop enclosed by dotted lines. Blocks G_2 & G_3 are in series, therefore their equivalent will be $(G_2 \cdot G_3)$.

In the feedback path block H_2 & $\frac{1}{G_3}$ are in series & the block H_1 is in parallel with this series combination.

$\therefore$ Their equivalent will be $\frac{H_2}{G_3}+H_1$

Step 3 : Now the internal loop is a feedback system with forward path T.F. equal to $(G_2 \, . \, G_3)$ & feedback path T.F. equal to $\left(H_1+\frac{H_2}{G_3}\right)$.

It is equivalent will be $\dfrac{G_2 . G_3}{1+G_2 G_3\left(H_1+\frac{H_2}{G_3}\right)}=\dfrac{G_2 . G_3}{1+G_2(H_2+H_1 G_3)}$

The B.D. reduces to

Step 4 : Now consider the internal loop enclosed by dotted lines. Two elements are in series in the forward path. This system is a positive feedback system.

$\therefore$ It is equivalent will be

$$\therefore \quad = \frac{G_1 . G_2 . G_3 / \{1+G_2(H_2+H_1 G_3)\}}{1-\dfrac{G_1 G_2 G_3}{1+G_2(H_2+H_1 G_3)} \times \dfrac{H_2}{G_3}}$$

$$= \frac{G_1 G_2 G_3}{1+G_2 H_2+G_2 . G_3 H_1-G_1 G_2 H_2}$$

$\therefore$ The B.D. will be

Step 5 : Now two blocks are in parallel. Therefore the T.F., $\frac{C}{R}$ will be

$$\frac{C}{R}=G_4+\frac{G_1 G_2 G_3}{1+G_2 H_2+G_2 G_3 H_1-G_1 G_2 H_2}$$

Example 34:

A C.S. is represented by the following B.D. Find the T.F., $\frac{C(s)}{R(s)}$

Solution:

In Block Diagram Reduction Technique we always try to simplify the inner loops first.

Step 1 : Shift the take off point of H_1 ahead of G_2. Hence for the internal loop, G_2 & G_3 will be in series & G_4 will be in parallel with + sign.

∴ The equivalent will be $(G_2\ G_3 + G_4)$ & B.D. reduces to

Step 2 : Shift the take-off point of H_1 i.e. (P) to (Q).

∴ B.D. will be

Step 3 : Now we have three clear cut feedback loops. Always solve the inner-loop first. Its equivalent will be

$$\frac{(G_2 . G_3 + G_4)}{1 + H_2 (G_2 G_3 + G_4)}$$

The B.D. becomes

Step 4 : Solve the next internal loop. The equivalent will be

$$\frac{G_1 [G_2 G_3 + G_4] / 1 + (G_2 G_3 + G_4) H_2}{1 + \dfrac{G_1 (G_2 G_3 + G_4)}{1 + (G_2 G_3 + G_4) H_2} \times \dfrac{H_1 G_2}{(G_2 - G_3 + G_4)}}$$

$$= \frac{G_1 (G_2 G_3 + G_4)}{1 + (G_2 G_3 + G_4) H_2 + G_1 G_2 H_1}$$

∴ B.D. reduces to

Step 5 : Now solve this feedback loop

[**Note :** When C(s) is directly feedback to the error detector, the feedback gain H(s) = 1, and the system is called unity feedback system]

$$\therefore \frac{C(s)}{R(s)} = \frac{G_1 (G_2 G_3 + G_4) / 1 + G_1 G_2 H_1 + H_2 (G_2 G_3 + G_4)}{1 + \dfrac{G_1 (G_2 G_3 + G_4)}{1 + G_1 G_2 H_1 + H_2 H_2 (G_2 G_3 + G_4)} \times 1}$$

$$\text{i.e. } \frac{C(s)}{R(s)} = \frac{G_1 G_2 G_3 + G_1 G_4}{1 + G_1 G_2 G_3 + G_1 G_4 + G_1 G_2 H_1 + G_2 G_3 H_2 + G_4 H_2}$$

5

Time Response Specifications

NTRODUCTION

The various terms associated with this time response are marked on the response; and they are as under.

Rise Time (t_r)

It is the time required by the response to raise from 10% to 90% of final value. For computation purpose it is taken as the time required by the response to reach to 100% of final value.

Derivation

$c(t) = 1$ at $t = t_r$. Substitute in equation (1)

$$\therefore \quad C(t_r) = 1 = 1 - \frac{e^{-\zeta\omega_n t_r}}{\sqrt{1-\zeta^2}}$$

$$\sin\left[\left(\omega_n \sqrt{1-\zeta^2}\right) t_r + \tan^{-1}\left(\frac{\sqrt{1-\zeta^2}}{\zeta}\right)\right] \quad \text{...(1)}$$

$$\therefore \quad \frac{e^{-\zeta\omega_n t_r}}{\sqrt{1-\zeta^2}}$$

$$\sin\left[\omega_n \sqrt{1-\zeta^2}\, t_r + \tan^{-1} \frac{\sqrt{1-\zeta^2}}{\zeta}\right] = 0$$

This can happen if

$$\sin\left[\omega_n \sqrt{1-\zeta^2}\, t_r + \tan^{-1} \frac{\sqrt{1-\zeta^2}}{\zeta}\right] = 0$$

i.e. $\omega_n \sqrt{1-\zeta^2}\, t_r + \tan^{-1}\frac{\sqrt{1-\zeta^2}}{\zeta} = \pi,$

2π ... [0 is not possible as it will give –ve value of t_r]

$$\therefore \omega_n \sqrt{1-\zeta^2}\, t_r + \tan^{-1}\frac{\sqrt{1-\zeta^2}}{\zeta} = \pi$$

$$\therefore t_r = \frac{\pi - \tan^{-1}\frac{\sqrt{1-\zeta^2}}{\zeta}}{\omega_n\sqrt{1-\zeta^2}} = \frac{\pi-\theta}{\omega_d} \quad ...(2)$$

After attaining the peak over shoot the response again attains 100% of the final value. For this time required will be

$\frac{2\pi-\theta}{\omega_d}$ and so on.

Significance : This term is the measure of speed of response of the system.

Peak time (t_p)

It is the time required by the response to reach the peak over shoot.

At $t = t_p$; $\frac{d\,c(t)}{dt} = 0$

$\therefore$ Differentiate with respect to time 't';

$$\therefore \frac{\sigma}{\sqrt{1-\zeta^2}} e^{-\sigma t_p} \sin[\omega_d t_p + \theta] -$$

$$\frac{e^{-\sigma t_p}}{\sqrt{1-\zeta^2}} \times \omega_d \cos[\omega_d t_p + \theta] = 0$$

$$\sigma \sin\left[\omega_d t_p + \theta\right] - \omega_d \cos\left[\omega_d t_p + \theta\right] = 0$$

i.e. $\zeta\omega_n \sin\left[\omega_d t_p + \theta\right] - \omega_n\sqrt{1-\zeta^2} \cos\left[\omega_d t_p + \theta\right] = 0$

from (6) $\zeta = \cos\theta$ & $\sqrt{1-\zeta^2} = \sin\theta$

$\cos\theta \sin[\omega_d t_p + \theta] - \sin\theta \cos[\omega_d t_p + \theta] = 0$

$\therefore \sin[\omega_d t_p + \theta - \theta] = 0$

$\therefore \omega_d t_p = \omega_n \sqrt{1-\zeta^2}\ t_p = \pi, 2\pi$ .... ['0' will not give practical result, hence not considered.]

$$\therefore \quad t_p = \frac{\pi}{\omega_n \sqrt{1-\zeta^2}} \qquad ...(3)$$

Similarly the second over shoot will occur at time; $\frac{2\pi}{\omega_n \sqrt{1-\zeta^2}}$ and so on.

The under root will occur at times,

$$\frac{3\pi/2}{\omega_n \sqrt{1-\zeta^2}}; \frac{5\pi/2}{\omega_n \sqrt{1-\zeta^2}} \text{ and so on.}$$

Significance : This term also is a measure of speed of response of the system.

Peak (Maximum) Over Shoot (M_p)

It is the largest deviation of the output over the step input during the transient state.

Derivation :

At $t = t_p$; $c(t) = 1 + M_p$

$\therefore$ Substitute $t = t_p = \frac{\pi}{\omega_n \sqrt{1-\zeta^2}}$

& $c(t) = 1 + M_p$ in equation (1)

$$\therefore \ 1+M_p = 1 - \frac{e^{\left(-\zeta\omega_n \times \frac{\pi}{\omega_n \sqrt{1-\zeta^2}}\right)}}{\sqrt{1-\zeta^2}} \sin\left[\omega_n \sqrt{1-\zeta^2} \times \frac{\pi}{\omega_n \sqrt{1-\zeta^2}} + \theta\right]$$

$$\therefore \ M_p = -\frac{e^{-\frac{\pi\zeta}{\sqrt{1-\zeta^2}}}}{\sqrt{1-\zeta^2}} \sin[\pi+\theta]$$

$$= -\frac{e^{-\frac{\pi\zeta}{\sqrt{1-\zeta^2}}}}{\sqrt{1-\zeta^2}} [-\sin\theta]$$

Also $\sqrt{1-\zeta^2} = \sin\theta$ From (eq. 4)

$$\therefore \quad M_p = e^{\frac{-\pi\zeta}{\sqrt{1-\zeta^2}}} \qquad ...(4)$$

Significance : M_p is function of ζ only. Hence it is the measure of relative stability of the system. As the value of ζ goes on increasing M_p goes on decreasing.

Settling time (t_s)

It is the time required by the response to decrease & stay within a specified percentage of its final value, usually 5% or 2%.

Derivation :

The envelop of the response consist of two terms; one sinusoidally oscillating [sin {$\omega_d t + \theta$}] & other exponentially decreasing $\left[\frac{e^{-\sigma t}}{\sqrt{1-\zeta^2}}\right]$. For such systems approximate expression for setting time is obtained when the envelop of sinusoidal term has entered the tolerance band, which is usually 2%. Hence for getting the equation only exponentially decaying envelop is considered

$$\therefore \quad \frac{e^{\zeta\omega_n t_s}}{\sqrt{1-\zeta^2}} = 0.02 \, ;$$

As ζ is small; $\sqrt{1-\zeta^2} \approx 1$

$$\therefore \quad e^{\zeta\omega_n t_s} = 0.02$$

$$\therefore \quad t_s = \frac{4}{\zeta\omega_n} \qquad ...(5)$$

For 5% tolerance band, $t_s = \frac{3}{\zeta\omega_n}$

Significance : This is the measure of speed of response of the system i.e. it indicates how fast system is going to settle within the tolerance band. From equation (5) it is clear that $t_s \propto \frac{1}{\zeta\omega_n}$ i.e. t_s is inversely proportional to the real part of root of characteristic equation. Hence as the location of roots [poles of closed - loop T.F.] shift away from j ω axis to the L.H.S. t_s reduces; hence speed of response improves.

For over damped system; $t_s = 4 \times$ Time constant for 2% accuracy

$$\therefore \frac{4}{\zeta\omega_n} = 4 \times T. \quad T = \frac{1}{\zeta\omega_n}$$

Delay time (t_d)

It is the time required by the response to reach 50% of its final value.

Steady state error (e_{ss})

It is the error between the actual & the desired output at $t \to \infty$. It is the specification of steady - state response. As it is 2nd order system,

$$e_{ss} = \frac{1}{Kv}$$

$$\therefore \quad e_{ss} = \frac{2\zeta}{\omega_n} \qquad ...(6)$$

Note :

(i) Out of these six terms, first five are measure of transient response & the sixth one i.e. e_{ss} is a measure of steady state response. The transient response consists of two parts :

(i) Speed of response; which is indicated by t_s and

(ii) Relative stability which is indicated by ζ, hence by M_p.

(ii) Equation (3) is,

$$\omega_n = \sqrt{\frac{Kv}{T}} \ \& \ \zeta = \frac{1}{2\sqrt{K_v T}}$$

$$\therefore \ Kv = \frac{\omega_n}{2\zeta} \ \& \ T = \frac{1}{2\zeta\omega_n} \qquad ...(7)$$

Hence $\zeta\omega_n$ is important term as it controls the rate of rise & decay of c(t).

Lower value of ζ increase Kv; hence reduces e_{ss}; therefore accuracy of the system will be better, But this will increase the time - constant hence it will slow - down the speed of the response. Higher value of ζ, hence increases e_{ss} but improves the speed of response. Therefore ζ is selected by considering all the necessary requirements. Most of the practical systems have ζ around 0.7.

TIME RESPONSE SPECIFICATIONS

In the previous section we have learned how to model mathematically the given physical system. This gives the relationship between input and output in terms of system parameters. Every control system has to perform certain task; i.e. we need its output as per the requirements. Time response is the evalution of system output response with respect to time and with reference to the input signal. The complete output response with respect to time is divided into two parts :

(i) Steady state response [C_{ss}]

(ii) Transient response [C_t (t)]

Hence the complete time response is

$$c(t) = C_{ss} + C_t (t)$$

Thus it is clear that for time - response analysis we need the input signals. So first we shall study the standard input test signals.

STANDARD TEST SIGNALS

Step Input

This signal represents an instantaneous change in the reference input variables.

Mathematically

$$\left.\begin{aligned} r(t) &= R; t \geq 0 \\ &= 0; t < 0 \end{aligned}\right\} \rightarrow \text{in time domain}$$

$$R(s) = \frac{R}{s} \rightarrow \text{in S - domain}$$

Where R is a constant.

This may also be written as

r(t) = R.u(t);

Where u(t) is a unit step function

Significance :

If input is the angular position of the shaft, then the step input represents sudden rotation of the shaft; hence sudden change in angular position of the shaft.

Advantage :

Its initial instantaneous jump in amplitude gives idea about system's quickness to respond.

Ramp Input

This signal represents a constant change in value with respect to time. It is the intergration of step function.

Mathematically;

$$r(t) = R.t \; ; \; t \geq 0$$

$$= 0 \; ; \; t < 0 \rightarrow \text{in time domain}$$

$$R(s) = \frac{R}{s^2} \rightarrow \text{in S - domain}$$

Where R is a constant.

This can also be written as

$$r(t) = R(t).\ u(t);$$

Where u(t) is a unit step function.

Significance :

If input is in the form of angular displacement of shaft, ramp input represents constant speed rotation of the shaft.

Advantage :

It has the ability to test how the system would respond to a signal which increases linearly with time.

Parabolic Input

This is the intergration of ramp input. Hence

$$\left.\begin{aligned} r(t) &= R.t^2 \geq 0 \\ &= 0; t < 0 \end{aligned}\right\} \text{ in time domain}$$

or $r(t) = R\ .\ t^2\ .\ u(t)$

$$R(s) = \frac{R}{s^3} \text{ in S - domain}$$

Significance : It is the fastest signal.

TIME RESPONSE OF A STANDARD 2ND ORDER UNDER DAMPED SYSTEM TO UNIT STEP INPUT

The standard form of 2nd order unity feedback system. T.F. is

$$T(s) = \frac{C(s)}{R(s)} = \frac{\omega_n^2}{s^2 + 2\zeta\omega_n s + \omega_n^2}$$

For unit step, input [R(s) = 1/s]

$$\therefore \quad c(s) = \left[\frac{\omega_n^2}{s^2 + 2\zeta\omega_n s + \omega_n^2}\right] \times \frac{1}{s}$$

The output response c(t) can be obtained by taking inverse Laplace Transform. Hence C(s) must be expressed in factored form.

$$\therefore\ s^2 + 2\zeta\omega_n s + \omega_n^2 =$$

$$= \left\{s+(-\zeta\omega_n + j\omega_n\sqrt{1-\zeta^2}\right\}\left\{s+(-\zeta\omega_n - j\omega_n\sqrt{1-\zeta^2})\right\}$$

$$\therefore\ C(s) = \frac{\omega_n^2}{s\left\{s+(-\zeta\omega_n + j\omega_n\sqrt{1-\zeta^2}\right\}\left\{s+(-\zeta\omega_n - j\omega_n\sqrt{1-\zeta^2})\right\}}$$

Therefore by method of residues

$$c(t) = \frac{\omega_n^2}{s^2+2\zeta\omega_n s+\omega_n^2}\Big|_{s=0} + 2\,\mathrm{Re}$$

$$\left[\frac{\omega_n^2}{s\{s+2\zeta\omega_n - j\omega_n\sqrt{1-\zeta^2}\}}\Big|_{s=\zeta\omega_n - j\omega_n\sqrt{1-\zeta^2}} \times e^{-\zeta\omega_n t}\, e^{-j\omega_n\sqrt{1-\zeta^2}\,t}\right]$$

$$\therefore\ c(t) = 1 - \frac{e^{-\zeta\omega t}}{\sqrt{1-\zeta^2}}\sin\left\{\omega_n\sqrt{1-\zeta^2} + \tan^{-1}\left(\frac{\sqrt{1-\zeta^2}}{\zeta}\right)\right\} \quad ...(1)$$

This is the equation for the output response of standard 2nd order unity feedback system when excited by a unit step input.

or $c(t) = 1 - \dfrac{e^{-\sigma t}}{\sqrt{1-\zeta^2}}\sin[\omega_d t + \theta]$

If the response 'c(t)' is plotted as a function of time 't'.

Impulse Signal

It is defined as the signal which has zero valve everywhere except at t = 0. At t = 0 its magnitude is infinite. Perfect impulse can not be achieved practically, therefore it is approximated by a pulse of small width and of unit area.

Mathematically

$$\delta(t) = 0;\ \text{for } t \neq 0$$

& $$\int_{-\varepsilon}^{+\varepsilon}\delta(t)\,dt = 1;\ \varepsilon \to 0$$

This function is the derivative of step function

$$\delta(t) = u(t)\ \&\ \pounds^{-1}[\delta(t)]$$

$$= 1 = R(s)$$

Note :

In many practical control system, the actual input may vary in random manner with respect to time; but for analysis and design purpose only these basic signals are considered, as these are simple to describe mathematically.

STEADY STATE RESPONSE (S.S.R)

It is that part of the system response which remains at $t \rightarrow \infty$

(Practically when time 't' becomes comparatively large)

Significance :

This response gives an indication of final accuracy of the system. If the S.S.R. does not agree with the desired response exactly, the system is said to have a **Steady State Error** (e_{ss}). Therefore from accuracy point of view e_{ss} should be minimum.

Steady State Error (e_{ss})

It is defined as the difference between actual system output and the desired system output as $t \rightarrow \infty$ [as 't' becomes large]. It is the measure of system accuracy i.e. system's state performance. There are two methods for evaluating the e_{ss} :

1. Error constant method
2. Error series method

Error Constant Method

In general the loop T.F.. G(s) . H(s) is given by

$$G(s).H(s) = \frac{K(1+T_1 s)(1+T_2 s)\ldots(1+T_m s)}{s^J (1+T_1{'}s')(1+T_2{'}s)\ldots(1+T_n{'}s)} \qquad \ldots(A)$$

$$= \frac{K'(s+z_1)(s+z_2)\ldots(s+z_m)}{s^J (s+p_1)(s+p_2)\ldots(s+p_n)} \qquad \ldots(B)$$

Equation (A) is the time constant form

K = Gain; T = time constant

Equation (B) is the pole zero form ;

z = zeros ; p = poles

(1) Type of C.S. :

Number of poles at the origin gives the type of the control system. Above equation represent type 'J' system. It means there are 'J' number of integrations in the system.

(2) Order of C.S. :

Total number of poles in the loop T.F. represents the order of the sytem. Above equations represent n^{th} order system.

(3) Determination of e_{ss} :

$$E(s) = R(s) - B(s)$$

$$= R(s) - H(s) . C(s)$$

The closed loop T.F. is

$$\frac{C(s)}{R(s)} = \frac{G(s)}{1+G(s).H(s)}$$

$$\therefore \; E(s) = R(s) - \frac{R(s).G(s).H(s)}{1+G(s).H(s)}$$

$$\therefore \; E(s) = \frac{R(s)}{1+G(s).H(s)} \quad ...(1)$$

For a unity feedback sytem H(s) = 0 ;

$$\therefore \quad B(s) = C(s)$$

In time domain b(t) = c(t)

$$e(t) = r(t) - c(t) \quad ...(2)$$

The steady state error is $e_{ss} = \lim_{t \to \infty} e(t)$

& By final value theorem $e_{ss} = \lim_{s \to 0} s . E(s)$ ...(3)

$$\therefore \; e_{ss} = \lim \frac{s.R(s)}{1+G(s).H(s)} \quad ...(4)$$

(4) Effect of input signal on e_{ss} :

(i) **Step input :** R(s) = R/s

Using equation (4)

$$\therefore \; e_{ss} = \lim_{s \to 0} \frac{R}{1+G(s).H(s)} = \frac{R}{1+\lim_{s \to 0} G(s).H(s)}$$

$$\therefore \; e_{ss} = \frac{R}{1+K_p}$$

Where $K_p = \lim_{s \to 0} G(s) . H(s)$

K_p is called as **step error - constant or position error constant.**

(ii) **Ramp input :** $R(s) = \frac{R}{s^2}$

Using equation (4)

$$\therefore\ e_{ss} = \lim_{s\to o} \frac{R}{s + sG(s).H(s)}$$

$$\therefore\ e_{ss} = \frac{R}{K_v}$$

Where $K_v = \lim_{s\to 0} s\,G(s)\,.\,H(s)$

K_v is called as **Ramp error Constant or velocity error constant.**

(iii) **Parabolic input :** $R(s) = \frac{R}{s^3}$

Using equation (4)

$$\therefore\ e_{ss} = \lim_{s\to 0} \frac{R}{s^2 + s^2\,G(s).H(s)} \qquad \therefore\ e_{ss} = \frac{R}{K_a}$$

Where $K_a = \lim_{s\to 0} s^2\,.\,G(s)\,.\,H(s)$

K_a is called as **parabolic error constant or acceleration error constant.**

(5) Effect of the type of system on e_{ss}:

From above derivations :

$K_p = \lim_{s\to 0} G(s)\,.\,H(s)$;

$K_v = \lim_{s\to 0} s\,G(s)\,.\,H(s)$

$K_a = \lim_{s\to 0} s^2\,G(s)\,H(s)$

Where G(s) H(s) is given by equation (A).

Substituting, different values of 'J'.

Type of the system (J)	K_p	K_v	K_a	Step input $e_{ss} = \frac{R}{1+K_p}$	Ramp Input $e_{ss} = \frac{R}{K_v}$	Parabolic input $e_{ss} = \frac{R}{K_a}$
0	K	0	0	R/1 + K	∞	∞
1	∞	K	0	0	R/K	∞
2	∞	∞	K	0	0	R/K
3	∞	∞	∞	0	0	0

Conclusions :

(i) Thus the steady state error 'e_{ss}' depends on the type of input signal, type of system (J) and the loop T.F. G(s) H(s).

(ii) Position, Velocity and Acceleration constant are significant only when the input signal is step, ramp and parabolic function respectively.

(iii) As the type of system (J) increases i.e. progressively more steady state errors are eliminated. But in practice system of type higher than 2 are not employed as their transient response is very poor and they are more difficult to stabilize.

(6) Draw backs of error constants :

(i) They do not give information about e_{ss} when inputs are other than the three basic signals.

(ii) They fail to indicate the exact manner in which error function changes with time. They only indicate the error as $t \to \infty$ and do not give it's value at any specific time.

Error Series Method

This method eliminates the drawbacks of above method.

$$E(s) = \left\{\frac{1}{1+G(s).H(s)}\right\} R(s) \ ;$$

$$W_e(s) = \frac{1}{1+G(s).H(s)}$$

The error series is defined as

$$e_s(t) = C_0 \,.\, r(t) + C_1\, r(t) + \frac{C_2}{2!}\, r(t) + ... + \frac{C_n}{n!}\, r^{[n]}(t) + ...\, r(t)$$

$$= \frac{d\,r(t)}{dt} \ ;\ r(t) = \frac{d^2\,r(t)}{dt^2}$$

C = error coefficients

$$C_0 = \lim_{s \to 0} W_e(s)$$

$$C_1 = \lim_{s \to 0} \frac{d\,W_e(s)}{ds} ; ....\, C_n = \lim_{s \to 0} \frac{d^n\,W_e(s)}{ds^n}$$

Note : In error series $e_s(t)$ rather than e_{ss} is obtained. e_{ss} represents steady state error i.e. the error as $t \to \infty$. From error series the error can be obtained at any time, hence $e_s(t)$ is the representation.

TRANSIENT RESPONSE $[C_T(T)]$

It is that part of the response which becomes zero as time becomes large $[t \rightarrow \infty]$. Hence this response exists for initial period only.

Significance :

It is important from dynamic behavior point of view. The nature of transient response depends on system poles. Hence on the type and order of the system. For analysing the transient response the step signal is considered as the input as

(i) Most of the practical signals are step signals in nature and

(ii) Analysis becomes simple.

Time Response of 1st Order System

The B.D. of 1st order system.

Wherein $G(s) = \dfrac{1}{T_s}$

$H(s) = 1$

$$T(s) = \frac{C(s)}{R(s)} = \text{closed – loop T.F.} = \frac{G(s)}{1+G(s).H(s)} = \frac{1}{1+Ts}$$

$$\therefore \quad C(s) = \frac{1}{1+T_s} \times R(s)$$

If R(s) is unit step input ;

i.e. $R(s) = \dfrac{1}{s}$

$$\therefore \quad C(s) = \frac{1}{s(1+T_s)}$$

$$\therefore \quad C(s) = \frac{1}{s} - \frac{T}{1+T_s}$$

Take inverse Laplace Transform

$\therefore\ c(t) = 1 - e^{-t/T}$

This is the output response of the system. The variation of output c(t) with time (t).

$$c(t) = 1 - e^{-t/T}$$

$$\therefore \quad \frac{d\,c(t)}{dt} = \frac{1}{T} e^{-t/T}$$

At t = 0; $\qquad$ c(t) = 0

& $\frac{d\,c(t)}{dt} = \frac{1}{T}$ = Slope of the response curve, figure 1.22

As $t \to \infty$; $c(t) \to 1$

Significance of T

T is called as **Time constant** of the system. It is the time at which output is 63% of input value.

i.e. at $t = T$; $c(t) = 1 - e^{-1} = 0.63$

'T' indicate how fast system reaches the final value; hence it indicates the behavior of the system before 't' becomes larges. Therefore it is indicative of transient response of the 1st order system. Small time constant gives the fast response.

For this 1st order system with unit step input; steady state error is

$$e_{ss} = \lim_{s\to 0} [s \,.\, E(s)] \qquad ...(1)$$

For a unity feedback system, the error signal is

$E(s) = R(s) - C(s)$

$$= \frac{1}{s} - \frac{1}{(Ts+1)s} = \frac{s(Ts+1)-s}{s^2(1+Ts)}$$

$$= \frac{Ts^2}{s^2(1+Ts)} = \frac{T}{1+Ts}$$

$$\therefore \quad e_{ss} = \lim_{s\to 0} [s \,.\, E(s)]$$

$$= \lim_{s\to 0}\left[s\left(\frac{T}{1+T_s}\right)\right] = 0 \qquad ...(2)$$

In T - domain;

$e(t) = r(t) - c(t) = 1 - \{1 - e^{-t/T}\} = e^{-t/T}$

$e_{ss} = \lim_{t\to\infty} [e(t)] = \lim_{t\to\infty} [e^{-t/T}] = 0$

Thus in a 1st order C.S. the system output tracks the unit step input with zero steady state error ($e_{ss} = 0$)

Time response of 2nd Order System

System can be represented by the B.D.

K_v = Gain T = Time constant

The closed loop T.F. = $T(s) = \frac{C(s)}{R(s)}$

$$= \frac{G(s)}{1+G(s).H(s)} = \frac{K_v/s(1+Ts)}{1+\frac{K_v}{s(1+Ts)}} \times 1$$

$$\therefore \frac{C(s)}{R(s)} = \frac{K_v}{Ts^2+s+K_v} = \frac{K_v/T}{s^2+\frac{1}{T}s+\frac{K_V}{T}}$$

$$\therefore \frac{C(s)}{R(s)} = \frac{K_v/T}{s^2+\frac{1}{T}s+\frac{K_V}{T}} \qquad ...(1)$$

The standard form of 2nd order system is

$$\therefore \frac{C(s)}{R(s)} = \frac{\omega_n^2}{s^2+2\zeta\omega_n s+\omega_n^2} \qquad ...(2)$$

Wherein $\omega_n^2 = \frac{K_v}{T}$ & $2\zeta\omega_n = \frac{1}{T}$

$$\therefore \omega_n = \sqrt{\frac{K_v}{T}} \qquad ...(3)$$

$$\zeta = \frac{1}{2\sqrt{K_v T}}$$

$\omega_n \rightarrow$ Un - damped natural frequency of oscillation

$\zeta \rightarrow$ Damping factor

Characteristic equation :

The characteristic equation of the system [q(s)] is defined as

$$q(s) = 1 + G(s)\ H(s)$$

Which is the denominator of the closed loop T.F.

It represents the poles of the T.F.

$$\therefore q(s) = 1 + G(s)\ .\ H(s) = s^2 + 2\ \zeta\ \omega_n\ s + \omega^2_n \qquad ...(4)$$

∴ The order of the characteristic equation will be same as the order of the system. Solution of characteristic equation [q(s) = 0] gives the location of the poles of the system.

STABILITY OF THE SYSTEM

A system is said to be stable if for bounded input, the output is also bounded.

Consider the characteristic equation of a 2nd order system

$$q(s) = s^2 + 2\ \zeta\ \omega_n\ s + \omega^2_n\ ;$$

The solution gives the location of poles i.e. roots of q(s) in S-plane

$$s = -\zeta\,\omega_n \pm j\,\omega_n\sqrt{1-\zeta^2}$$

$$s_1 s_2 = -\sigma \pm j\omega_d \qquad \text{...(5)}$$

wherein $\sigma = \zeta\omega_n$; $\quad \omega_d = \omega_n\sqrt{1-\zeta^2}$

ω_d is called as the damped natural frequency of oscillation. With the help of above equation (5); the roots of characteristics equation (which represents the poles) can be located on the S-plane. The S-plane is as shown in figure (2.9) wherein σ and $-\sigma$ axes are called as real axes whereas $j\,\omega_d$ and $-j\,\omega_d$ are called as imaginary axes.

Effect of *z* on Root - location & System Response

(A) For damping factor $\zeta < 0$ (i.e. if ζ is negative)

s_1 & s_2 (roots of characteristic equation or poles). will have +ve real part; and may be complex or real.

These roots are located on the right hand side of imaginary axis (RHS of S-plane)

Such systems are called as *negatively damped* system and they show *unstable* response.

(B) For damping factor $\zeta = 0$

s_1 & s_2 will have only complex parts i.e. s_1 & $s_2 = \pm\, j\,\omega_n$.

∴ Roots are located on the imaginary axis. Such system show *sustained oscillations* and are called as *un-damped* system.

(C) For damping factor $0 < \zeta < 1$

s_1 & s_2 will be having –ve real parts and roots will be complex conjugates. Roots will be located on LHS of $j\omega$ – axis. Such system show damped *oscillatory response* and are called as *under - damped* systems.

(D) For damping factor $\zeta = 1$

s_1 & $s_2 = -\,\omega_n$ i.e both the roots will be same and are located on the –ve real axis. Such system show exponentially increasing response; but bounded one and are called as *critically damped* systems.

(E) For damping factor $\zeta > 1$

s_1 & s_2 will have only real values, but – ve. Hence both the roots will be located on LHS of $j\,\omega$ – axis. Such systems also show exponentially increasing response ; but bounded one ; and are called as *over damped* system.

Thus by observing the output responses for various values of ζ following conclusions can be drawn :

(i) A closed system is stable if all the poles i.e. roots of characteristic equation lie on the LHS of j ω axis of S-plane.

(ii) As the value of ζ is increased from 0 towards 1, root location moves from imaginary axis towards –ve real. At $\zeta = 1$, the roots meet on –ve real axis. If ζ is further increased one root moves towards origin and other towards infinity of the –ve real axis.

(iii) For time response analysis, $\zeta > 0$

Note :

Most of the practical C.S. have ζ between 0 and 1 i.e. they are under damped systems; hence for time response analysis under damped system is considered. Most practical systems have $\zeta = 0.7$. For such systems, s_1 & s_2 are complex conjugates and may be located as under :

The line joining the origin & s_1 called as ζ – line.

The angle of ζ – line with respect to –ve real axis is denoted by θ.

Now distance $[0 - s_1] = \sqrt{(\zeta\omega_n)^2 + \{\omega_n\sqrt{1-\zeta^2}\}^2} = \omega_n$

$$\therefore \quad \cos\theta = \zeta\frac{\omega_n}{\omega_n} = \zeta$$

$$\therefore \quad \theta = \cos^{-1}\zeta.$$

Also, $$\sin\theta = \omega_n\frac{\sqrt{1-\zeta^2}}{\omega_n} = \sqrt{1-\zeta^2}$$

$$\& \quad \tan\theta = \frac{\sqrt{1-\zeta^2}}{\zeta} \qquad \text{...(6)}$$

SOLVED EXAMPLES

Example 1:

Without controller determine % M_p and t_s for a unit step input.

For $k_D = \frac{1}{30}$, a PD controller is introduced in forward path (in series) with $\frac{100}{s(s+12)}$ Find new values of ζ, M_p, t_s.

Solution:

Without Controller

$$\frac{C(s)}{R(s)} = \frac{100}{s^2 + 12s + 100}$$

$$\omega_n^2 = 100 \text{ or } \omega_n = 10$$

$$\therefore 2\zeta\omega_n = 12$$

$$\therefore \quad \zeta = 0.6$$

$$M_p = e^{\frac{-\pi\zeta}{\sqrt{1-\zeta^2}}} = 0.094 = 9.4\ \%$$

$$t_s = \frac{4}{\zeta\omega_n} = 0.66 \text{ sec}$$

The controller part is

$$1 + \frac{s}{30} = \frac{s+30}{30}$$

$$G(s) = \frac{(100)}{s(s+12)} \times \frac{(s+30)}{30} = \frac{3.33(s+30)}{s(s+12)}$$

$$\therefore \frac{C(s)}{R(s)} = \frac{G}{1+GH}, \ H(s) = 1$$

$$= \frac{3.33(s+30)}{s^2 + 12s + 3.33(s+30)} = \frac{3.33s + 100}{s^2 + 15.33s + 100}$$

$$\omega_n^2 = 100 \qquad \therefore \omega_n = 10$$

$$2\zeta\omega_n = 15.33$$

$$\therefore \zeta = \frac{15.33}{2\omega_n} = 0.766 \text{ s}$$

$$\omega_d = \omega_n \sqrt{1-\zeta^2} = 6.4224 \text{ rad/s} \left(\frac{-\pi\zeta}{\sqrt{1-y^2}}\right)$$

$$M_p = 100 \times e = 2.353\ \%$$

$$t_s = \frac{4}{\zeta\omega_n} = 0.5218 \text{ sec}$$

Example 2:

Show that the system transfer function Y(s)/X(s) has a zero in the right half s-plane Obtain y(t) when x(t) is a unit step for the system shown in figure below.

Solution:

$$y(t) = -1 - 3e^{-2t} + 4e^{-t}$$

Example 3:

For a servomechanism represented by

$$\frac{d^2\theta}{dt^2} + 10\frac{d\theta}{dt} - 150e = 0$$

Where e is the actuating signal, calculate the damping ratio, natural frequency and damped frequency of oscillations.

Solution:

$\zeta = 0.41$

$\omega_n = 12.25$ rad/sec

$\omega_d = 11.17$ rad/sec

Example 4:

A system is described by

$$\frac{d^2y}{dt^2} + 8\frac{dy}{dt} + 25\,y(t) = 50\,x(t)$$

Evaluate the response and maximum output for a step of 2.5 units.

Solution:

$$y(t) = 5\,[1 - 1.66\,e^{-4t} \sin (3t + 37^\circ)]; \quad y_{max} = 5.07$$

Example 5:

Measurements conducted on a servomechanism show the error response to be

$$e(t) = 1.66\,e^{-8t} \sin (6t + 37^\circ)$$

Where the input is given a sudden unit displacement. Determine the natural frequency, damping ratio and damped angular frequency.

Solution:

$\omega_n = 9.96$ rad/sec ;

$\zeta = 0.8; \quad \omega_d = 6$ rad/sec

Example 6:

In the block diagram shown in fig. below.

(G)s = A/s^2 and H(s) = ms + n. For A = 20, determine the values of m and n for a step input with a time constant 0.1 sec which gives a peak oveshoot of 30%.

Solution :

m = 2; n = 78.07

Example 7:

Find the step ramp and parabolic error coefficients for unity feedback systems which have the following transfer functions.

(i) $\dfrac{50}{(1+0.1\,s)\,(1+2s)}$

(ii) $\dfrac{10\,(s+3)}{(s+1)\,(s^2+2s+2)}$

(iii) $\dfrac{4\,(s^2+10\,s+100)}{s^2\,(s+3)\,(s^2+6s+10)}$

(iv) $\dfrac{K}{s(s+3)\,(s+25)}$

Solution:

(i) $K_p = 50$; $K_v = K_a = 0$;

(ii) $K_p = 15$; $K_v = K_a = 0$

(iii) $K_p = \infty = K_v$; $K_a = 40/3$

(iv) $K_p = \infty$; $K_v = K/75$; $K_a = 0$

Example 8:

A unity feedback system has

$$G(s) = \frac{K\,(2s+1)}{s\,(4s+1)\,(s+1)^2}$$

Determine the value of K if the steady-state value of error is to be less than 0.1 when an input r(t) = 1 + 5t is applied.

Solution:

K > 50

Example 9:

The open loop transfer function of a control system with unity feedback is given by

$$G(s) = \frac{150}{s(1+0.25\,s)}$$

(a) Evaluate the generalised error series for the system.

(b) Determine the steady-state error for an input $r(t) = (1 + t^2)\,u(t)$

Solution:

(a) $e(t) = 0 + \dfrac{2t}{150} + ...$

(b) $e_{ss} = \infty$.

Example 10:

For a unity feedback C.S. the open-loop T.F. is

$$G(s) = \frac{K[s+2]}{s[s^3+7\,s^2+12s]}$$

(i) State the type and order of the system

(ii) Find K_p, K_v & K_a

(iii) Find e_{ss} due to step, ramp and parabolic inputs

(iv) Find e_{ss} due to an input described by

$$r(t) = 1+2\,t+\frac{3\,t^2}{2}.$$

Solution:

(i) The type and order of the system is determined from the loop T.F. G(s) . H(s)

$$G(s).H(s) = \frac{K(s+2)}{s[s^3+7\,s^2+12\,s]}\times 1$$

$$\therefore\ G(s).H(s) = \frac{K(s+2)}{s^2[s^2+7\,s+12]}$$

Type = No. of poles at origin

$\therefore$ Type = 2

Order = No. of Poles in the system

∴ Order = 4

∴ It is 4th order type – 2 system.

(ii) Determination of error - constant

$$K_p = \lim_{s\to 0} G(s).H(s)$$

$$\therefore\ K_p = \lim_{s\to 0}\left\{\frac{K(s+2)}{s^2(s^2+7s+12)}\right\} = \infty$$

$$K_v = \lim_{s\to 0} s.G(s).H(s)$$

$$K_v = \lim_{s\to 0}\left[s\times\left\{\frac{K(s+2)}{s^2(s^2+7s+12)}\right\}\right] = \infty$$

$$K_a = \lim_{s\to 0} s^2 G(s).H(s)$$

$$K_a = \lim_{s\to 0}\left[s^2\times\left\{\frac{K(s+2)}{s^2(s^2+7s+12)}\right\}\right]$$

$$\therefore\ K_a = \frac{K}{6}$$

(iii) Determination of e_{ss}

For Step input, $e_{ss} = \dfrac{R}{1+K_P}$

∴ $e_{ss} = 0$

For Ramp input, $e_{ss} = \dfrac{R}{K_v}$

∴ $e_{ss} = 0$

For parabolic input; $e_{ss} = \dfrac{R}{K_a}$

$$e_{ss} = \frac{R}{K/6} = \frac{6.R}{K}\ \{\text{if } R = 1\}$$

(iv) At input $r(t) = 1 + 2t + \dfrac{3}{2}t^2$.

Taking Laplace Transform

$$\therefore\ R(s) = \frac{1}{s}+\frac{2}{s^2}+\frac{3}{s^3}$$

= Unit step input + 2 × Unit ramp input + 3 × Unit parabolic input.

$$\therefore\ e_{ss} = e_{ss}|_{Step} + 2\times e_{ss}|_{Ramp} + 3\times e_{ss}|_{Parabolic}$$

$$= 0 + 0 + 3\left\{\frac{6}{K}\right\}$$

$$\therefore \; e_{ss} = \frac{18}{K}$$

Example 11:

The open-loop T.F. of a unity feedback system is $G(s) = \frac{9}{s+1}$. Using the generalized error series, determine the e_{ss} when the system is excited by following inputs.

(i) r(t) = 2;

(ii) r(t) = t,

(iii) $r(t) = \frac{3t^2}{2}$ &

(iv) $r(t) = 1 + 2t + \frac{3}{2}t^2$

Also find the error after 4 seconds in each case.

Solution:

The error series is,

$$e_s(t) = C_0 r(t) + \frac{C_1 . \dot{r}(t)}{1!} + \frac{C_2 . \ddot{r}(t)}{2!} + \ldots$$

Where $C_0 = \lim_{s\to 0} W_e(s)$ &

$$C_n = \lim_{s\to 0} \frac{d^n [W_e(s)]}{ds^n}$$

$$W_e(s) = \frac{1}{1+G(s).H(s)} = \text{Error T.F.}$$

$$\therefore \; C_0 = \lim_{s\to 0}\left[\frac{1}{1+\frac{9}{s+1}}\right] = 0.1$$

$$C_1 = \lim_{s\to 0}\left\{\frac{d}{ds}\left(\frac{1}{1+9/s+1}\right)\right\}$$

$$= \lim_{s\to 0}\left[\frac{d}{ds}\left(\frac{s+1}{s+10}\right)\right]$$

$$= \lim_{s \to 0} \left[\frac{9}{(s+10)^2} \right] = 0.09$$

$$C_2 = \lim_{s \to 0} \left\{ \frac{d^2}{ds^2} \left(\frac{1}{1 + \frac{9}{s+1}} \right) \right\} = -0.018$$

As max. order or input is 2; coefficients C_3, C_4, ... will not be required.

(i) $r(t) = 2$

$\therefore \quad \dot{r}(t) = 0$

$\therefore \; e_s(t) = C_0 \; r(t) = 0.1 \times 2 = 0.2$

$\therefore$ The error $e_s(t)$ is constant i.e. it is not time dependent. Hence after 4 sec. also it will remain 0.2

(ii) $r(t) = t$

$\therefore \quad \dot{r}(t) = 1 \quad \& \quad \ddot{r}(t) = 0$

$$\therefore \quad e_s(t) = C_0 \, r(t) + \frac{C_1 \, \dot{r}(t)}{1!}$$

$$= 0.1 \times t + \frac{0.09 \times 1}{1} = 0.1 \; t + 0.09$$

$\therefore$ After 4 sec,

$e_s(4) = 0.1 \times 4 + 0.09 = 0.49$

(iii) $r(t) = \frac{3}{2} t^2$

$\therefore \quad \dot{r}(t) = 3 \; t$

$\quad \ddot{r}(t) = 3 \; \& \; \dddot{r}(t) = 0$

$$\therefore \quad e_s(t) = C_0 \, r(t) + \frac{C_1 \, . \, \dot{r}(t)}{1!} + \frac{C_2 \, \ddot{r}(t)}{2!}$$

$$= 0.1 \times \frac{3}{2} t^2 + 0.09 \times 3t + (-0.018) \times \frac{3}{2 \times 1}$$

$$= 0.15 \; t^2 + 0.27 \; t - 0.027$$

$\therefore$ After 4 sec,

$e_s(4) = 0.15 \times (4)^2 + 0.27 \times 4 - 0.027 = 3.453$

(iv) $\qquad r(t) = 1 = 2 \; t + \frac{3}{2} \; t^2$

$$\therefore \quad \dot{r}(t) = 2 + 3\,t$$

$$\ddot{r}(t) = 3 \quad \dddot{r}(t) = 0$$

$$\therefore \; e_s(t) = 0.1\left[1 + 2t + \frac{3}{2}t^2\right] + 0.09\,[2 + 3\,t] - 0.018\,\frac{[3]}{2}$$

$$= 0.1 + 0.21\,t + 0.15\,t^2 + 0.18 + 0.27\,t - 0.027$$

$$\therefore \; e_s(t) = 0.15\,t^2 + 0.47\,t + 0.253$$

$$\therefore \; e_s(4) = 0.15\,(4)^2 + 0.47 \times 4 + 0.253 = 4.533$$

Example 12:

Find error coefficients of Figure below.

Solution:

The feedback blocks are in parallel, see Figure.

$$H(s) = \frac{s^3\,(s+3)+10}{s^3\,(s+5)} = \frac{s^4 + 3\,s^3 + 10}{(s)^3\,(s+5)}$$

$$\therefore \; G(s)H(s) = \left(\frac{s}{s+10}\right)\left(\frac{s^4 + 3\,s^3 + 10}{s^3\,(s+5)}\right)$$

$$= \frac{1}{s^2}\left(\frac{s^4 + 3\,s^3 + 10}{(s+10)\,(s+5)}\right)$$ It is a type 2 system.

Hence $k_p = k_v = \infty$ and k_a is finite.

$$k_p = \lim_{s \to 0} G(s)\,H(s)$$

$$= \lim_{s \to 0} \frac{1}{s^2}\left(\frac{s^4 + 3\,s^3 + 10}{(s+10)\,(s+5)}\right) = \infty$$

$$k_v = \lim_{s \to 0} s\,G(s)H(s)$$

$$= \lim_{s \to 0} \frac{1}{s}\left(\frac{s^4 + 3\,s^3 + 10}{(s+10)\,(s+5)}\right) = \infty$$

$$k_a = \lim_{s \to 0} s^2\,G(s)H(s)$$

$$= \lim_{s \to 0}\left(\frac{s^4 + 3\,s^3 + 10}{(s+10)\,(s+5)}\right)$$

$$= \left(\frac{0+0+0}{(10)\,(5)}\right) = \frac{1}{5} = 0.2$$

Example 13:

For a system $G(s)H(s) = \dfrac{k}{s^2(s+2)(s+3)}$

Find k to limit the error for input

$1 + 10\,t + \dfrac{40}{2}\,t^2$ to 10.

Solution:

$$R(s) = \pounds\left[1 + 10\,t + \frac{40}{2}\,t^2\right]$$

$$= \frac{1}{s} + \frac{10}{s} + \frac{40}{s^3} = \frac{(s^2+10s+40)}{s^3}$$

$$e_{ss} = \lim_{s\to 0} \frac{s.R(s)}{1+G(s)H(s)}$$

$$= \lim_{s\to 0} \frac{s.\left(\dfrac{s^2+10+40}{s^3}\right)}{1+\dfrac{k}{s^2(s+2)(s+3)}}$$

$$= \lim_{s\to 0} \frac{\left(s^2+10s+40\right)}{1+\dfrac{k}{(s+2)(s+3)}} \quad = \frac{k}{0+\dfrac{k}{(2)(3)}}$$

$$= \frac{40}{\dfrac{k}{6}} = \frac{240}{k}$$

But $e_{ss} = 10$

$\therefore \quad 10 = \dfrac{240}{k}$ or k = 24

Example 14:

For r(t) = 0.1 t, it is desired that $e_{ss} \le 0.005$. Find the value of k for which this takes place. Refer figure below.

Solution:

$$G(s) = \frac{k}{s(s+1)}\ ;\ H(s) = 1$$

$$\therefore \; G(s)H(s) = \frac{k}{s(s+1)}$$

As a type 1 system, it will have a finite error for a ramp input.

$$R(s) = £\,[r\,(t)] = £\,[0.1\ t] = \frac{0.1}{s^2}$$

$$\therefore \; e_{ss} = \lim_{s\to 0}\left\{\frac{s\,R(s)}{1+G(s)H(s)}\right\} = \lim_{s\to 0}\left\{\frac{\frac{0.1}{s^2}\times s}{1+\frac{k}{s(s+1)}}\right\}$$

$$= \lim_{s\to 0}\frac{0.1}{s+\frac{k}{s+1}} = \frac{0.1}{k}$$

$$\therefore \; e_{ss} \le 0.005 \quad \text{i.e.} \quad \frac{0.1}{k} \le 0.005$$

$$\therefore \; \frac{0.1}{0.005} \le k \quad \therefore \quad k \ge 20$$

Example 15:

For a unity feedback system :

$$G(s) = \frac{200}{s(s+8)}$$

$$r(t) = 2\ t$$

Find steady state error. Find value of k to reduce the error by 5%

Solution:

$$G(s) = \frac{200}{s(s+8)}$$

$$R(s) = £\ r(t) = \frac{2}{s^2}$$

$$\therefore \; e_{ss} = \lim_{s\to 0}\frac{s\,R(s)}{1+G(s)H(s)} = \lim_{s\to 0}\frac{s\frac{2}{s^2}}{1+\frac{200}{s(s+8)}}$$

$$e_{ss} = \lim_{s\to 0}\frac{2}{s+\frac{200}{s+8}} = \frac{2}{200/8} = 0.08$$

Now new error is 5% less

i.e. $e_{newss} = 0.95 \cdot e_{ss}$

$= 0.95 \times 0.08 = 0.76$

$$e_{ss} = \lim_{s \to 0} \frac{2}{s + \frac{k}{s+8}} = \frac{2}{k/8} = \frac{16}{k}$$

$$\therefore \frac{16}{k} = 0.076$$

$$\text{or} \quad k = \frac{16}{0.076} = 210.52$$

Example 16:

A unity feedback C.S. has forward path T.F. $G(s) = \frac{K}{s(s+10)}$. Determine the gain K so that the system has damping ratio of 0.5. For this value of K determine rise time (t_r), peak time (t_p) peak over shoot (M_p), setting time (t_s), steady state error (e_{ss}) and no. of oscillations system sustains before settling [N_s]. Also draw the approximate response indicating all the specifications.

Solution:

$$G(s) = \frac{K}{s(s+10)}, \; H(s) = 1$$

$\therefore$ q(s) = characteristic equation = 1 + G(s) . H(s)

$$= 1 + \frac{K}{s(s+10)} \times 1 = 0$$

$$\therefore q(s) = s^2 + 10\,s + k = 0$$

Equate with the std, form of 2nd order q(s) i.e.

$$s^2 + 2\,\zeta\,\omega_n s + \omega^2_n = 0$$

$$\therefore 2\,\zeta\,\omega_n = 10 \;\&\; \omega^2_n = K$$

Required $\zeta = 0.5$

$$\therefore 2 \times 0.5 \times \omega_n = 10$$

$$\therefore \quad \omega_n = 10$$

$$\therefore \quad K = \omega^2_n = 100$$

Determination of Time Response Specifications :

(i) Rise Time (t_r) :

$$t_r = \frac{\pi - \tan^{-1}\left(\frac{\sqrt{1-\zeta^2}}{\zeta}\right)}{\omega_n\sqrt{1-\zeta^2}} = \frac{\pi - \tan^{-1}\left\{\frac{\sqrt{1-0.25}}{0.5}\right\}}{10\sqrt{1-0.25}}$$

$\therefore\ t_r = 0.242$ sec

(ii) Peak time (t_p) :

$$t_p = \frac{\pi}{\omega_n\sqrt{1-\zeta^2}} = \frac{\pi}{10\sqrt{1-0.25}} \qquad \therefore\ t_p = 0.362 \text{ Sec}$$

(iii) Peak Over - shoot (M_p) :

$$M_p = \left(\frac{-\pi\zeta}{e^{\sqrt{1-\zeta^2}}}\right) = e^{\frac{-\pi\times 0.5}{\sqrt{1-0.25}}} \qquad \therefore\ M_p = 0.162 = 16.2\%$$

(iv) Settling time (t_s) :

$$t_s = \frac{4}{\zeta w_n} = \frac{4}{0.5\times 10}$$

$\therefore\ t_s = 0.8$ Sec

(v) No. of oscillations before setting (N_s):

Let f is the frequency in Hz then

$N_s = f \times t_s \qquad \omega_d = 2\pi f$

$$\therefore\ f = \frac{\omega_d}{2\pi}$$

$$\therefore\ N_s = \frac{\omega_d \times t_s}{2\pi} = \omega_n\frac{\sqrt{1-\zeta^2}}{2\pi}\times\frac{4}{\zeta\omega_n}$$

$$\therefore\ N_s = \frac{2\sqrt{1-\zeta^2}}{\pi\zeta}$$

$$\therefore\ N_s = 2\frac{\sqrt{1-0.25}}{\pi\times 0.5} = 1.1$$

$\therefore\ N_s = 1.1$

(vi) Steady state Error (e_{ss}) :

$$e_{ss} = \frac{2\zeta}{\omega_n} = \frac{2\times 0.5}{10}$$

$\therefore\ e_{ss} = 0.1 = 10\%$

Approximate Response will be as under

Example 17:

The system in the diagram consist of a unity feedback loop & a rate feedback loop.

(a) Without rate feedback determine the damping ratio, natural frequency, over shoot, t_s and the steady state error to a unit ramp input.

(b) Determine the rate feedback constant which will increase the equivalent damping factor of the system to 0.8. Determine the overshoot of the system and e_{ss} for unit ramp input

Solution:

Without rate feedback

(K = 0). Characteristic equaition is 1 + G(s) H(s) = 0

i.e. $1+\frac{16}{s(s+4)}\times 1 = s^2 + 4s + 16 = 0$

$\therefore \quad \omega_n^2 = 16$ i.e. $\omega_n = 4$ rad/sec

$2\omega_n \zeta = 4$ i.e. $\zeta = 0.5$

& $M_p = e^{-\pi\zeta/\sqrt{1-\zeta^2}} = 16.4\%$

& $t_s = \frac{4}{\zeta\omega_n} = 2$ sec

$$e_{ss} = \frac{2\zeta}{\omega_n} = 0.25$$

With rate feedback,

$\therefore$ Characteristic equation $1+\frac{16}{s(s+4)}\times[1+K_s] = 0$

$$s^2 + (4 + 16K)s + 16 = 0$$

$\therefore \quad \omega_n = 4$ Required

$$\zeta = 0.8$$

$\therefore (4 + 16K) = 2 \times 0.8 \times 4 = 6.4$

$\therefore 16K = 2.4$

i.e. $K = \frac{24}{16} = \frac{0.3}{2} = 0.15$

$\therefore K = 0.15 \quad t_s = \frac{4}{\zeta\omega_n} = \frac{4}{3.2} = 1.25$ Sec

$M_p = 6.8\%$ $e_{ss} = \dfrac{2\zeta}{\omega_n} = 0.4$

Example 18:

A feedback system employing output rate damping is shown in figure below.

(i) In absence of derivative feedback ($K_d = 0$) determine the damping factor and natural frequency of the system. What is the steady state error resulting out of unit step and unit ramp inputs ?

(ii) Determine K_d which will increase damping factor to 0.6. Calculate also the steady state error to ramp input with this value of K_d.

(iii) What value of K_A and K_d will yield the damping factor of 0.6 while maintaining the steady state error at 0.2 for unit ramp input ?

Solution:

(i) $K_d = 0$

$$\therefore\ G(s) = \frac{10}{s(s+2)};\ H(s) = 1$$

$q(s) = 1 + G(s) \cdot H(s)$

$= s^2 + 2\zeta\omega_n s + \omega_n^2 = 0 \rightarrow$ Standard form

For the given system

$$q(s) = 1 + \frac{10}{s(s+2)} \times 1 = s^2 + 2s + 10 = 0$$

Equating with the standard form

$2\zeta\omega_n = 2;\ \omega_n^2 = 10$

$\therefore\quad \omega_n = 3.16\ \&\ \zeta = 0.316$

It is type one system;

$\therefore\ e_{ss}$ for step input $= 0$

For ramp input; $e_{ss} = \dfrac{2\zeta}{\omega_n} = \dfrac{2 \times 0.316}{3.16} = 0.2$

$\therefore\ e_{ss} = 0$ for unit step input

$= 0.2$ For unit ramp input

(ii) K_d is introduced

$\therefore$ Internal feedback path can be simplified as $\dfrac{1/s(s+2)}{1 + \dfrac{1}{s(s+2)} \times K_d \cdot s}$

$$= \frac{1}{s^2 + (2+K_d)s}$$

$$\therefore\ G(s) = \frac{10}{s[s+(2+K_d)]} \ \&\ H(s) = 1$$

$$\therefore\ q(s) = 1 + \frac{10}{s[s+(2+K_d)]} \times 1 = 0$$

$\therefore\ q(s) = s^2 + (2 + K_d)\ s + 10 = 0$

Equating with the std. form

$\omega^2_n = 10; \qquad \therefore\ \omega_n = 3.16$

& $(2 + K_d) = 2\ \zeta\ \omega_n$.

Required $\zeta = 0.6$

$\therefore\quad 2 + K_d = 2 \times 0.6 \times 3.16$

$\therefore\quad K_d = 1.8$

e_{ss} for ramp input $= \dfrac{2\zeta}{\omega_n} = \dfrac{2 \times 0.6}{3.16}$

$\therefore\quad e_{ss} = 0.38$

Note : ζ is increased from 0.316 to 0.6 due to rate feedback hence transient response is improved; but e_{ss} has increased from 0.2 to 0.38; hence S.S.R. has degenerated.

(iii) Now to preserve the S.S.R. ($e_{ss} = 0.2$) obtaining the desired transient response ($\zeta = 0.6$); K_A & K_d are being modified:

$e_{ss} = \dfrac{2\zeta}{\omega_n} = 0.2 \rightarrow$ Required

$\zeta = 0.6 \rightarrow$ Required

$\therefore\quad \dfrac{2 \times 0.6}{\omega_n} = 0.2$

$\therefore\quad \omega_n = 6$

Now $(2 + K_d) = 2\ \zeta\ \omega_n$.

$\therefore\quad K_d = 2 \times 0.6 \times 6 - 2$

$\therefore\quad K_d = 5.2$

Also $K_A = \omega^2_n$. $\qquad \therefore\quad K_A = 36$

Example 19:

A simple closed-loop system has a dynamic response defined by relation

$$T_m \frac{d^2 \theta_0}{dt^2} + \frac{d\theta_0}{dt} = K_v (\theta - \theta_0)$$

Where T_m is the motor time constant and K_v is the velocity constant. When this system is subjected to a unit step input $\theta_1(t) = u(t)$, the response obtained is given by

$\theta_0(t) = 1 - 1.25\ e^{-4.245\ t}\ [\sin (5.66\ t + 0.927)]$

Determine the values of

(i) Damping ratio

(ii) Peak overshoot

(iii) Setting time

(iv) No. of oscillations during setting time

(v) Steady state torque

Solution:

The system given is a 2nd order system the response of this system to a unit step input is

$$c(t) = 1 - \frac{e^{-\zeta\omega_n t}}{\sqrt{1-\zeta^2}}$$

$$\sin\left[\omega_n \sqrt{1-\zeta^2}\ t + \tan^{-1}\frac{\sqrt{1-\zeta^2}}{\zeta}\right] \quad ...(1)$$

Equating this with the given response i.e.

$\theta_0(t) = 1 - 1.25\ e^{-4.245\ t} \sin$

$[5.66\ t + 0.927]$ ...(2)

(i) $\therefore \quad \dfrac{1}{\sqrt{1-\zeta^2}} = 1.25$

$$\therefore \quad 1.25\sqrt{1-\zeta^2} = 1$$

$$\therefore\ 1.56\ [1 - \zeta^2] = 1$$

i.e. $\quad 0.56 = 1.56\ \zeta^2.$

$$\therefore \quad \zeta = \sqrt{\frac{0.56}{1.56}} = \pm 0.6$$

As the response is under - damped $0 < \zeta < 1$ $\quad \therefore\ \zeta = 0.6$

(ii) $M_p = e^{\frac{-\pi\zeta}{\sqrt{1-\zeta^2}}}$

$\therefore \quad M_p = 0.094 = 9.4\ \%$

(iii) $t_s = \dfrac{4}{\zeta\omega_n}$;

Equating (1) & (2) for $e^{-\zeta\omega_n t}$ & $e^{-4.245t}$

$\zeta\,\omega_n = 4.245$

$\therefore\ \omega_n = 7.075$

& $t_s = 0.942$ Sec

(iv) $N_s = 2\dfrac{\sqrt{1-\zeta^2}}{\pi\zeta}$

$\therefore \quad N_s = 0.85$

(v) Steady State torque = The output which exists in the steady state condition

$$= e_{ss} = \frac{2\zeta}{\omega_n} \qquad\qquad \therefore\ e_{ss} = 0.1696$$

Example 20:

(a) For the control system in Figure below

Find out the values of a and b such that the system will have the ratio of resonant frequency to the undamped natural frequency equal to 0.707 and a time constant equal to 0.10 sec. Take A=10.

Solution:

$G(s) = \dfrac{A}{s(s+10)}$ $H(s) = (a\,s + b)$

$\therefore\ q(s) = 1 + G(s)\,.\,H(s)$

$$= 1 + \frac{10}{s(s+10)} \times (a\,s + b) = 0$$

$\therefore\ s^2 + (10 + 10\,a)\,s + 10\,b = 0$

Equating with the std. 2nd order equation

i.e. $s^2 + 2\,\zeta\,\omega_n s + \omega^2_n = 0$

$2\,\zeta\,\omega_n = 10 + 10\,a$; $\omega^2_n = 10\,b$ Given,

$$\frac{\text{Resonant Frequency } \omega_d}{\text{Natural Frequency } \omega_n} = 0.707 \qquad ...(1)$$

& T = 0.10 ...(2)

(1) gives $\omega_n \frac{\sqrt{1-\zeta^2}}{\omega_n} = 0.707$

$\therefore \zeta = 0.707$

(2) gives $\frac{1}{T} = 2\zeta\omega_n = \frac{1}{0.1}$

$\therefore \omega_n = 7.07$

We have $\omega_n^2 = 10\ b$

$\therefore b = 5$

& $2\ \zeta\ \omega_n = 10 = 10 + 10\ a$

$\therefore a = 0$

Example 21:

Unit step response data of a unity feed back second order closed loop system is given below:

time in sec	output
0	0
0.05	0.25
0.10	0.80
0.15	1.08
0.20	1.12
0.25	1.02
0.30	0.98
0.35	0.98
0.40	1.00
0.45	1.00
0.50	1.00

Obtain

(i) Peak Over shoot

(ii) Peak time

(iii) Settling time

(iv) Damping factor

(v) Natural frequency of oscillations

(vi) Output Response equation

(vii) Open - Loop T.F.

(viii) Closed Loop T.F.

Solution:

Plot the response on graph paper

From the graph following results may be concluded :

(i) Peak over - shoot occurs at t = 0.2 sec.

$\therefore t_p = 0.2$ sec

(ii) Peak over - shoot is $M_p = 0.12$ i.e. 12%

(iii) System settles within 2% tolerance band after 0.3 sec.

$\therefore$ Settling time; $t_s = 0.3$ sec

(iv) Damping factor; $M_p = e^{-\pi\zeta/\sqrt{1-\zeta}}$ $\therefore \zeta = 0.56$

(v) Natural frequency of Oscillation : $t_s = \dfrac{4}{\zeta\omega_n}$;

$$\therefore 0.3 = \frac{4}{0.56 \times \omega_n}$$

$\therefore \omega_n = 23.83$ rad/sec

(vi) $$c(t) = 1 - \frac{e^{-\zeta\omega_n t}}{\sqrt{1-\zeta^2}} \sin\left[\omega_n \sqrt{1-\zeta^2}\, t + \tan^{-1}\frac{\sqrt{1-\zeta^2}}{\zeta}\right]$$

$\therefore c(t) = 1 - 0.82\, e^{-13.32\,t} \sin[19.75\,t + 0.95]$

(vii) Open loop T.F.

$$G(s) = \frac{\omega_n^2}{s(s+2\zeta\omega_n)}$$

$$\therefore G(s) = \frac{568}{s(s+28.35)}$$

(viii) Closed - Loop T.F.

$$\frac{c(s)}{R(s)} = \frac{\omega_n^2}{s^2+2\zeta\omega_n s+\omega_n^2} \quad \therefore \frac{C(s)}{R(s)} = \frac{568}{s^2+28.35s+568}$$

Example 22:

Assuming system in above example 49 is simulated using a series R-L-C circuit. Obtain the values of R,L and C to obtain to given data. Assume that input to the circuit is terminal Voltage from a battery.

Solution:

For R-L-C series circuit;

$$\frac{C(s)}{R(s)} = \frac{1/LC}{s^2 + \frac{R}{L}s + \frac{1}{LC}}$$

Equating with the T.F.

$$\therefore \quad \frac{R}{L} = 28.35 \qquad \text{...(i)}$$

$$\& \quad \frac{1}{LC} = 568 \qquad \text{...(ii)}$$

There are three unknowns & two equations. Hence one the variables should be fixed first.

Let C = 1 μ f

∴ from (ii)

L = 1760 H

Put in (i) R = 49.91 k Ω.

Example 23:

For the C.S. shown in the figure obtain the characteristics equation & calculate the following when unit step input is applied to the system:

(i) ω_n

(ii) ζ

(iii) M_p

(iv) t_s

(v) Time at which max. & min. will occur

(vi) Equation of the output

Solution:

$$G(s) = \frac{20 \times 1.2}{s(s+1)(s+5) \times 0.2} = \frac{120}{s(s+1)(s+5)} \;\&\; H(s) = \frac{s}{6}$$

$$\therefore q(s) = 1 + G(s) \cdot H(s) = 1 + \frac{120}{s(s+1)(s+15)} \times \frac{s}{6} = 0$$

$$= (s+1)(s+5) + 20 = 0$$

$$\therefore q(s) = s^2 + 6s + 25 = 0$$

Equating with the std form i.e.

$$q(s) = s^2 + 2\,\zeta\,\omega_n s + \omega_n^2 = 0$$

(i) $\omega_n^2 = 25$ $\therefore\ \omega_n = 5$ rad/sec

(ii) $2\,\zeta\,\omega_n = 6$ $\therefore\ \zeta = 0.6$

(iii) $M_p = e^{-\frac{\pi\zeta}{\sqrt{1-\zeta^2}}}$

$\therefore$ $M_p = 9.4\ \%$

(iv) $t_s = \dfrac{4}{\zeta\omega_n}$ $\therefore\ t_s = 1.33$ sec

(v) t_p = time for over - shoot (Max)

$$= \frac{\pi}{\omega_n\sqrt{1-\zeta^2}} = 0.785 \text{ sec}$$

t_u = time for under shoot (Min)

$$= \frac{2\pi}{\omega_n\sqrt{1-\zeta^2}} = 1.575 \text{ sec}$$

(vi) Equation of the output $c(t) = 1 - \dfrac{e^{-\zeta\omega_n t}}{\sqrt{1-\zeta^2}}$

$$\sin\left[\omega_n\sqrt{1-\zeta^2} + \tan^{-1}\frac{\sqrt{1-\zeta^2}}{\zeta}\right]$$

$$\therefore \qquad c(t) = 1 - \frac{e^{-3t}}{0.8}\sin\left[4 + \tan^{-1}1.33\right]$$

Example 24:

Measurement conducted on a servo mechanism gives the system response as

$c(t) = 1 + 0.2\ e^{-60t} - 1.2\ e^{-10t}$ when subjected to a unit step input

(i) Obtain the expression for the closed loop T.F.

(ii) Determine the natural frequency & damping ratio of the system.

Solution:

The response has only exponentially decaying terms. Oscillating term is not present. Hence it is clear that system response is not under damped. As system is stable; response will be either critically damped or over damped.

$\therefore \zeta \geq 1$

(i) $c(t) = 1 + 0.2\, e^{-60t} - 1.2\, e^{-10t}$

Take L.T.

$$\therefore\ c(s) = \frac{1}{s} + \frac{0.2}{s+60} - \frac{1.2}{s+10}$$

Simplification gives

$$\therefore\ C(s) = \frac{600}{s(s+60)(s+10)}$$

$$\text{i.e.}\quad C(s) = \frac{600}{s(s^2+70s+600)}$$

$$R(s) = \frac{1}{s}$$

$$\therefore\ \frac{C(s)}{R(s)} = \frac{600}{s(s^2+70s+600)} \times s$$

$$\therefore \frac{C(s)}{R(s)} = \text{Closed loop T.F.} = \frac{600}{s^2+70s+600}$$

(ii) Equating with the std. form, $\dfrac{\omega_n^2}{s^2+2\zeta\omega_n s+\omega_n^2}$

$\omega_n^2 = 600$

$\therefore \omega_n$ - 24.5 rad/sec

& $2\zeta\omega_n = 70$ $\quad\therefore \zeta = 1.43$

Example 25:

The open loop transfer function of a unity feedback system is given by

$$G(s) = \frac{K}{s(1+T_s)}$$

Where K and T are positive constants. By what factor should the amplifier gain be reduced so that the peak overshoot of unit step response of the system is reduced from 75% to 25%.

Solution:

$G(s) = \dfrac{K}{s(1+T_s)}$ and $H(s) = 1$

$\therefore q(s) = 1 + G(s) \,.\, H(s) = 1 + \dfrac{K}{s(1+T_s)} \times 1 = 0$

$\therefore\ q(s) = Ts^2 + s + K = 0$

i.e. $q(s) = s^2 + \frac{1}{T} + \frac{K}{T} = 0$

Equating with the standard form

$$2\zeta\omega_n = \frac{1}{T} \ \&\ \omega_n^2 = \frac{K}{T}$$

$$\therefore \quad \zeta = \frac{1}{2\sqrt{KT}} \quad ...(1)$$

$$M_p = e^{-\pi\zeta/\sqrt{1-\zeta^2}}$$

Let $0.75 = e^{-\frac{\pi\zeta_1}{\sqrt{1-\zeta_1^2}}}$

$$\therefore \quad \frac{\zeta_1}{\sqrt{1-\zeta_1^2}} = 0.09$$

i.e. $\frac{\zeta_1^2}{1-\zeta_1^2} = 0.0084$

$$\therefore\ \zeta_1^2 = 0.0083 \quad ...(2)$$

Similarly Let $0.25 = e^{\frac{-\pi\zeta_2}{\sqrt{1-\zeta_2^2}}}$

$$\therefore\ \zeta_2^2 = 0.163 \quad ...(3)$$

Let K_1 is gain for M_p = 75% i.e. for ζ_1.

& K_2 is gain for M_p = 25% i.e. for ζ_2.

From (1), $\frac{\zeta_2^2}{\zeta_1^2} = \frac{4K_1T}{4K_2T}$

i.e. $\frac{\zeta_2^2}{\zeta_1^2} = \frac{K_1}{K_2}$

Using (2) & (3)

$$\frac{K_1}{K_2} = \frac{0.163}{0.0083}$$

$\therefore\ K_1 = 19.56\ K_2$.

$\therefore\ K_2 = 0.0509\ K_1$.

$\therefore$ Gain must be reduced by factor 0.0509.

Example 26:

The system equation is

$$\frac{d^2 y}{dt^2}+4\frac{dy}{dt}+8y = 8x$$

Find the Time domain specification for unit step input. (y → output x → input)

Solution:

$$\frac{d^2 y}{dt^2}+4\frac{dy}{dt}+8y = 8x$$

∴ Taking Laplace for zero initial conditions

$$s^2 Y(s) + 4\,s\,Y(s) + 8\,Y(s) = 8\,X(s)$$

$$\therefore (s^2 + 4s + 8)\,Y(s) = 8\,X(s)$$

$$\therefore \frac{Y(s)}{X(s)} = \frac{8}{s^2+4s+8}$$

Comparing with standard form

$$\frac{C(s)}{R(s)} = \frac{\omega_n^2}{s^2+2\zeta\omega_n s+\omega_n^2}$$

$$\omega_n^2 \rightarrow 8$$

$$\therefore \omega_n = \sqrt{8} = 2.828 \text{ rad/s}$$

$$2\,\zeta\,\omega_n \rightarrow 4 \qquad \therefore \zeta = \frac{4}{2\sqrt{8}} = 0.707 \text{ rad/s}$$

$$\omega_d = \omega_n\sqrt{1-\zeta^2} = 2.828\sqrt{1-(0.707)^2}$$

$$= 2 \text{ rad/sec}$$

Specifications

(1) $$t_d = \frac{1+0.7\zeta}{\omega_n} = \frac{1+0.7\left(\frac{1}{\sqrt{2}}\right)}{\sqrt{8}} = 0.5285$$

(2) $$t_r = \frac{\pi-\beta}{\omega_n}\quad \beta = \tan^{-1}\frac{\sqrt{1-\zeta^2}}{\zeta} = 0.7853$$

$$= \frac{\pi - 0.7853}{\omega_d} = 1.178 \text{ sec}$$

(3) $$t_p = \frac{\pi}{\omega_d} = 1.5707 \text{ sec}$$

(4) $$\% M_p = 100 \times e^{-\frac{\zeta\pi}{\sqrt{1-\zeta^2}}} = 4.321\%$$

(5) $$t_s = \frac{4}{\zeta\omega_n} = \frac{4}{\left(\frac{1}{\sqrt{2}}\right)\times\sqrt{8}} = 2 \text{ sec}$$

$$c(t) = 1 - \frac{e^{-\zeta\omega_n t}}{\sqrt{1-\zeta^2}} \sin(\omega_d t + \theta)$$

$$c(t) = [1 - 1.41\ e^{-2t} \sin(2t + 0.7853)]$$

Example 27:

A feedback system, Find k_1 & k_2 so that peak overshoot of the system is 1.2% and corresponding time is 1 sec.

Solution:

The closed loop transfer function standard form is

$$\frac{C(s)}{R(s)} = \frac{\omega_n^2}{s^2 + 2\zeta\omega_n s + \omega_n^2}$$

for our problem

$$\frac{C(s)}{R(s)} = \frac{G}{1+GH} = \frac{\frac{k_1}{s(1+s)}}{1+\frac{k_1}{s(1+s)}(1+k_2 s)}$$

$$= \frac{k_1}{s(1+s)+k_1(1+k_2 s)} = \frac{k_1}{s^2+(k_1 k_2+1)s+k_1}$$

$$\omega_n^2 \to k_1 \text{ or } \omega_n = \sqrt{k_1}$$

$2\ \zeta\ \omega_n \to 1 + k_1\ k_2.$

or $$\zeta = \frac{1+k_1 k_2}{2\omega_n} = \frac{1+k_1 k_2}{2\sqrt{k_1}}$$

$$\omega_d = \omega_n\sqrt{1-\zeta^2} = \sqrt{k_1}\sqrt{1-\frac{(1+k_1 k_2)^2}{4k_1}}$$

$$\therefore \%M_p = 100 \times e^{-\frac{\zeta\pi}{\sqrt{1-\zeta^2}}}$$

$$\therefore \quad 0.012 = e^{-\frac{\zeta\pi}{\sqrt{1-\zeta^2}}}$$

Taking natural log

$$\text{Ln }(0.012) = -\frac{\zeta\pi}{\sqrt{1-\zeta^2}}$$

$$4.4228 = \frac{\zeta\pi}{\sqrt{1-\zeta^2}}$$

$$1.4078 = \frac{\zeta}{\sqrt{1-\zeta^2}}$$

Squaring $1.9819\,(1-\zeta^2) = \zeta^2$.

or $\zeta = 0.8152$

Also $T_p = \frac{\pi}{\omega_d} = 1$

$\therefore \omega_d = \pi$.

But $\zeta = \frac{1+k_1 k_2}{2\sqrt{k_1}} = 0.8152$ ...(1)

$$\omega_d = \sqrt{k_1} = \sqrt{1-\frac{(1+k_1 k_2)^2}{4k_1}} = \pi$$

From (1) $1 + k_1k_2 = 1.63\sqrt{k_1}$.

Put in (2)

$$\omega_d = \sqrt{k_1} \times \sqrt{1-\frac{(163)^2 k_1}{4k_1}} = \sqrt{k_1}\sqrt{1-\frac{2.657}{4}}$$

$= \sqrt{k_1}\ 0.5794$ but $\omega_d = \pi$.

$$\therefore k_1 = \frac{\pi^2}{0.5294^2} = 29.40$$

From (1) $1 + k_1k_2 = 1.63\sqrt{k_1}$

$\therefore k_1k_2 = 7.838$

$$\therefore k_2 = \frac{7.838}{k_1} = 0.2665$$

Example 28:

Find k & k_1 so that $\zeta = 0.6$ and $t_s = 0.1$.

Solution:

First find out $\frac{C}{R}$ using block diagram reduction,

$$\therefore \text{ For inner loop } \frac{\frac{100}{1+0.25}}{1+\frac{100}{1+0.2\,s}k_t}$$

$$= \frac{100}{1+0.2\,s+100\,k_t}$$

$$\text{Overall } G(s) = \frac{5\,k}{s[1+100\,k_t+0.2\,s]}$$

$$\text{T.F.} = \frac{G}{1+GH} = \frac{5k}{0.2\,s^2+s(100\,k_t+1)+5\,k}$$

$$= \frac{25k}{s^2+5s(100\,k_t+1)+25\,k}$$

Comparing with standard function $\omega_n^2 = 25\text{ k}$

or $\omega_n = 5\sqrt{k}$

$2\,\zeta\,\omega_n = 5\,(1 + 100\,k_t)$

$$\text{or } \zeta = \frac{5(1+100\,k_t)}{10\sqrt{k}}$$

$\zeta = 0.6$ and $t_s = 0.1$

$$\text{Now } t_s = \frac{4}{\zeta\omega_n}\quad 0.1 = \frac{4}{0.6\times\omega_n}$$

$\therefore\ \omega_n = 66.67$ rad/sec; $\quad \omega_n = 5\sqrt{k} \qquad \therefore\ k = 177.78$

$$\text{From } \zeta = \frac{5(1+100\,k_t)}{10\sqrt{k}} = 0.6$$

$k_t = 0.118$

Example 29:

An R-L-C series circuit is an example of second order function. If R = 1 Ω, L = 1 H, C = 1 F. Find step response for a step voltage of 1V connected as input. Output is across R.

Solution:

Find the Transfer function and then output for given input.

The circuit in s-domain is

$$v_0 = 1 \,.\, I(s)$$

$$v_1 = I(s)\left[s+\frac{1}{s}+1\right]$$

$$I_1(s) = \frac{v_1 \,.\, s}{(s^2+s+1)}$$

$$\therefore\; v_0 = \frac{v_1 s}{s^2+s+1}$$

$$\frac{v_0}{v_1} = \frac{s}{s^2+s+1}$$

When v_1 = step

$$\therefore\; v_1(s) = \frac{1}{s}$$

$$\therefore\; v_0 = \frac{s}{s^2+s+1}.\, v_1(s) = \frac{1}{s^2+s+1}$$

$$\therefore\; v_0(t) = L^{-1}\frac{1}{s^2+s+1}$$

Complete denominator for perfect square.

$$v_0(t) = L^{-1}\frac{1}{s^{2+}\;s+\frac{1}{4}+\frac{3}{4}}$$

$$= L^{-1}\frac{1}{\left(s+\frac{1}{2}\right)^2+\frac{3}{4}}$$

$$= L^{-1}\frac{1}{\left(s+\frac{1}{2}\right)^2+\left(\frac{\sqrt{3}}{2}\right)^2}$$

$$v_0(t) = e^{-1/2\,t}\sin\frac{\sqrt{3}}{2}t$$

6

Electrical Potential and Circuits

INTRODUCTION

The electric field is a force field so far the charges are concerned, there as work associated with the movement of the charges in an electric field. If a test charge is moved along the direction of field, work is done by the field. Since the force exerted by the field on the charge is in the direction moved against the direction of the field, an external agent has to supply the energy to over come the force exerted on the charge by the field. This force is opposite to the direction of movement of the charge.

Consider the displacement of test charge q by an infinitesimal distance $d\bar{l}$ from A to B at an angle with the electric field $\bar{E}$ at a point A is shown in Fig. 1. The force exerted on the test charge by the field has magnitude qE and is directed along E. Its component along the line from A to B is qE cos α. If the charge is moved from A to B, the amount of work done dw by the field is the product of force and displacement.

$$dw = qE \cos \alpha \; d\bar{l}$$

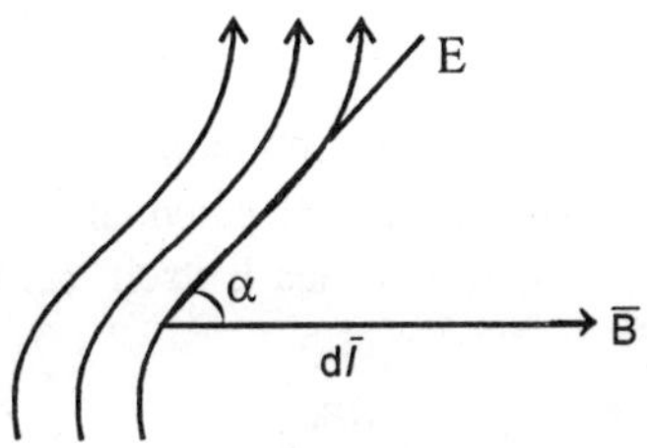

Fig. 1

$$= q\bar{E}.d\bar{l}$$

Where $d\bar{l}$ is the vector from A to B.

The work done W_{AB} by the field is moving a test charge q from A to B along a given path can be obtained by dividing the path into several segments of infinitesimal length all.

The result is a line integral expression given by

$$W_{AB} = q\int_A^B \bar{E}.d\bar{l}$$

The test charge has certain potential energy associated with it by virtue of its location in the electric field. W_{AB} as given by the above equation is then the loss of potential energy associated with the movement of the charge from A to B. Dividing W_{AB} by q gives the potential energy per unit charge. This quantity denoted by V_{AB} is known as the potential difference between the points A & B. Thus,

$$V_{AB} = \frac{W_{AB}}{q} = \int_A^B \bar{E}.d\bar{l}$$

If V_{AB} is positive, there is a potential energy associated with the movement of the charge from A to B, that is, the field does the work. If V_{AB} is negative, there is a gain is potential at energy associated with the moment of the charge from A to B, that is on external agent has to do the work.

ELECTRIC POTENTIAL DUE TO A CHARGE

If point A is taken to be at infinity & potential at infinity is taken to be zero, the potential V at a point B will be

$$V = \int_\infty^B E.d\bar{l}$$

$$V = -\int_\infty^r \frac{q}{4\pi\varepsilon_r\varepsilon_0 r^2} d_r \; ; = -\frac{q}{4\pi\varepsilon_r\varepsilon_0}\left[\frac{1}{r}\right]_\infty^r$$

$$= \frac{q}{4\pi\varepsilon_r\varepsilon_0 r}.$$

Since field intensity is the variation of potential with the distance, it can also be visualised as potential gradient. If potential for all points of space are known, the components of E and thus E itself can be found by taking the following derivatives.

$$E_x = \frac{-\partial V}{\partial x},\; E_y = \frac{-\partial V}{\partial y};\; E_z = \frac{-\partial V}{\partial z}$$

Therefore,

$$\bar{E} = \left(\bar{\imath}\,\frac{\partial V}{\partial x} + \bar{j}\,\frac{\partial V}{\partial y} + \bar{k}\,\frac{\partial V}{\partial z}\right)$$

Thus the unit of electric field is volt/meter.

POTENTIAL DUE TO GROUP OF CHARGES

The potential at a point due to a group of point charges q_1, q_2, q_n, is the algebraic sum of the potentials due to each charge. That is

$$V = \frac{1}{4\pi\varepsilon_0}\left(\frac{q_1}{r_1} + \frac{q_2}{r_2} + \dots \frac{q_n}{r_n}\right)$$

$$= \frac{1}{4\pi\varepsilon_0}\cdot\sum\cdot\frac{q_n}{r_n}$$

If the charge distribution is continuous, then the potential at a point is given by

$$V = \frac{1}{4\pi\varepsilon_0}\int\frac{dq}{r}$$

Where dq is a infinitesimal element of charge at a distance r from the point.

POTENTIAL DUE TO AN ELECTRIC DIPOLE

The potential at a point P due to the dipole shown in fig. will be

$$V = \frac{1}{4\pi\varepsilon_0}\left(\frac{q}{r_1} - \frac{q}{r_2}\right)$$

$$= \frac{q}{4\pi\varepsilon_0}\cdot\left(\frac{r_2 - r_1}{r_2 r_1}\right)$$

Now if $r \geq 2a$, $\theta = \alpha$

Therefore $(r_2 - r_1) = 2a \cos\alpha$

$\cong 2a \cos\theta$

and $r_1 r_2 \cong r^2$

Therefore $V = \dfrac{q2a\cos\theta}{4\pi\varepsilon_0 r^2}$

$$= \frac{1}{4\pi\varepsilon_0}\cdot\frac{\rho\cos\theta}{r^2}$$

ELECTRIC POTENTIAL ENERGY

For two charges $+q_1$ and $-q_2$ placed at a distance apart, energy is stored is the system because a definite amount of work has to be done to move change into kinetic energy and as a result, they will accelerate towards each other.

Thus potential energy of a system of point charges is the work required to assemble these charges by bringing them together from infinity.

Now, the potential due to q_1 at q_2 location B

$$V = \frac{1}{4\pi\varepsilon_0} \cdot \frac{q_1}{r}.$$

The work done required to more q_2 from infinity to distance r by definition of potential will be

$$w = Vq_2$$

$\therefore$ the electric potential energy $V_1 = Vq_2$

$$= \frac{1}{4\pi\varepsilon_0} \cdot \frac{q_1 q_2}{r}.$$

MAXWELL'S EQUATION FOR STATIC FIELD

Since the potential difference between two points A & B is independent of the path followed from A to B. Therefore for two paths ACB & ADB as shown in Fig. 2.

$$\int_{ACB} \overline{E}.d\bar{l} = \int_{ADB} \overline{E}.d\bar{l}$$

or $\int_{ACB} \overline{E}.d\bar{l} - \int_{ADB} \overline{E}.d\bar{l} = 0$

or $\int_{ACB} \overline{E}.d\bar{l} + \int_{BDA} \overline{E}.d\bar{l} = 0$

or $\oint_{ACB} \overline{E}.d\bar{l} = 0$

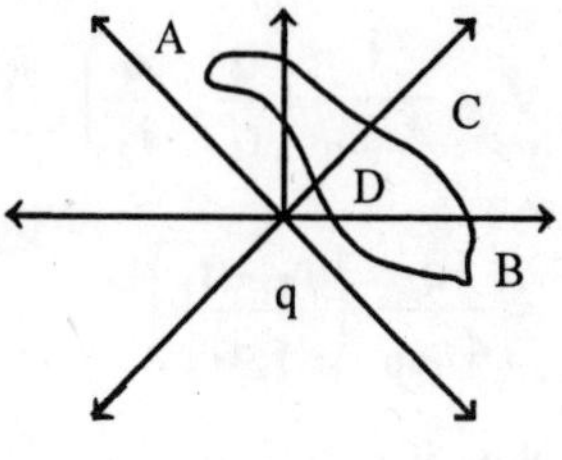

Fig. 2

The integral is taken around a closed path.

By superposition the above result can be generated for discrete or continuous charges as

$$\oint \overline{E}.d\bar{l} = 0$$

This simply implies that the work involved in moving a test charge around a closed path in a static electric field is equal to zero. In differential from, this can be written as :

$$\nabla \times \overline{E} = 0.$$

GROUPING OF CELLS

Cells can be grouped either in series or in parallel as shown in Fig. 3. In series grouping, the current flowing through the external circuit of n cells connected in series is given by

$$i = \frac{nE}{R + nr}$$

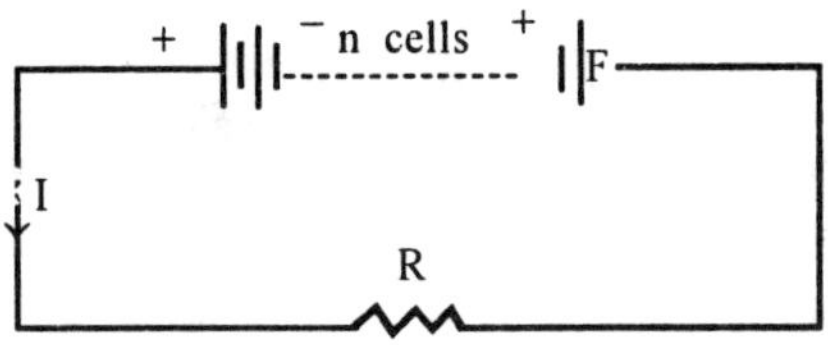

Fig. 3

Where r is internal resistance of a single cell.

In the parallel grouping shown in Fig. 4 the current is

$$i = \frac{E}{R + r/n}$$

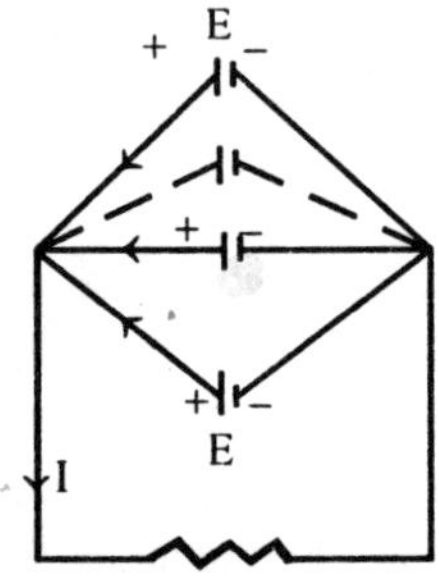

Fig. 4

SECONDARY CELL OR STORAGE CELL

The action of secondary cell is essentially the same as that of primary cells except that the secondary cells are charged by supplying electrical energy, the electrical energy supplied is converted into chemical energy and then stored. The charged secondary cells are capable of supplying electrical energy during discharging, the chemical energy stored is converted into electrical energy.

THE CONCEPT OF CAPACITANCE

Consider a conducting sphere of radius R with charge q and no conducting or non conducting bodies nearby. Its potential will be

$$V_+ = \frac{1}{4\pi\varepsilon o} - \frac{q}{R}$$

Another similar sphere with charge – q and at a large distance from first sphere will have potential of

$$V_- = \frac{1}{4\pi\varepsilon o} - \frac{q}{R}$$

The potential difference between the two electrically isolated spheres will be

$$V = V_+ - V_- = \frac{1}{4\pi\varepsilon o} - \frac{2q}{R}$$

This means that the potential difference is proportional to the charge q of each sphere

Alternatively,

$$q = (2\pi\varepsilon_0 R)\, V = CV$$

Where C is called the capacitance of the system of two sphere.

Now when the two spheres are moved close together the spherical symmetry of their lines of force will be spoiled with the result that the potential of positive sphere lowers while that of negative sphere will be raised. This has an importance implication i.e. though the charges on the spheres remaining the same, the potential difference has reduced. Thus the capacitance C = Q/V has been made larger than the situation in which spheres were at a large distance.

From the above concepts, capacitance of a single isolated conductor can be defined by imagining a second conductor carrying equal and opposite charge and of infinite radius centered about the main conductor. Clearly, the potential to this infinitely distant sphere will be zero. Then the capacitance of the conductor of radius R will be

$$C = \frac{Q}{V} = 4\pi\varepsilon oR$$

Unit of capacitance is

Farad = Coulomb/Volt.

It is a very large unit and therefore micro farad (uf) and picofarad (pf) are used.

$$I\mu F = 10^{-6}\,F$$

and $1pF = 1^{-12}F$

When the above concepts are generalised in terms of two nearby conductors, called plates of any shape and carrying equal and opposite charges the system is called a capacitor the capacitance of a conductor depends on the geometry of each plate, their spatial relationship and the medium between them. This arrangement is shown in Fig. 5.

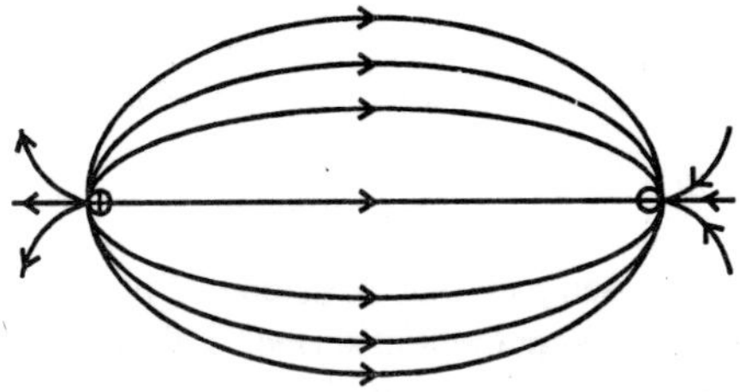

Fig. 5

Capacitors can be used to establish electric fields between the plates and thus store electric energy in small volumes .

Pallel Plate Capacitors

It is formed by two parallel conducting plates of area A, separated by a distance d and charged to +q and -q respectively. It d is small as compared to the length of the plates, the electric field will be uniform with in the plates, and the fringing of lines of force at the edges of plates can be neglected.

Consider a gaussian surface of height h and of area equal to plate area A is shown in Fig. 6

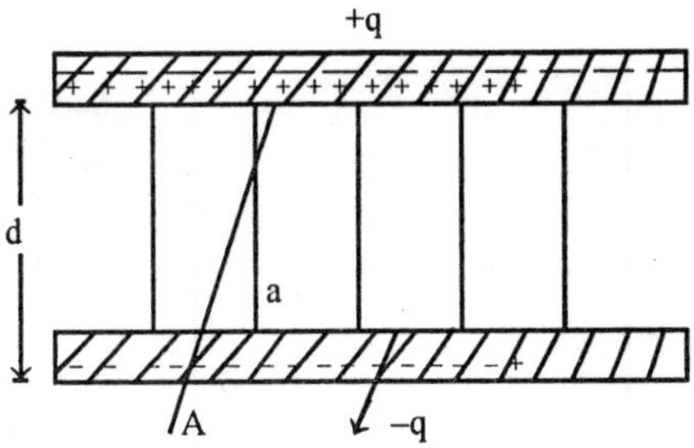

Fig. 6

For the part of gaussion surface lying inside the capacitor plate, the flux due to E is zero because the electric field inside a conductor carrying a static charge is zero. Moreover the flux of E through the vertical surfaces of guassian surface is also zero.

For the face of the gaussian surface that lies between the plates E is constant and flux ϕ_B is simply EA. Therefore by guass's law

$$\epsilon_0 \oint \bar{E} . d\bar{s} = q$$

or $\epsilon_0 EA = q$

The potential difference between the two plates is

$$V = -\int_a^b E . dl$$

Where a-b is a chosen path that starts on one plate and ends on the other because each plate is an equipotential surface & the electrostatic force is path independent. Therefore, V = Ed and

$$C = \frac{q}{V} = \frac{\epsilon_0 EA}{Ed} = \frac{\epsilon_0 A}{d}$$

CAPACITOR WITH A DIELECTRIC

When a dielectric material is placed in a uniform electric field existing in a capacitor, the field tends to separate the negative and positive charge in the molecules. Thus electric dipoles are induced in the material which tend to align along the direction of the field. The result is that the centre of positive charge of the material separates from the centre of negative charge and though the material remaining electrically neutral it becomes polarised. In other words, surface charges +q' and -q' are induced on the dielectric as shown in Fig. 7.

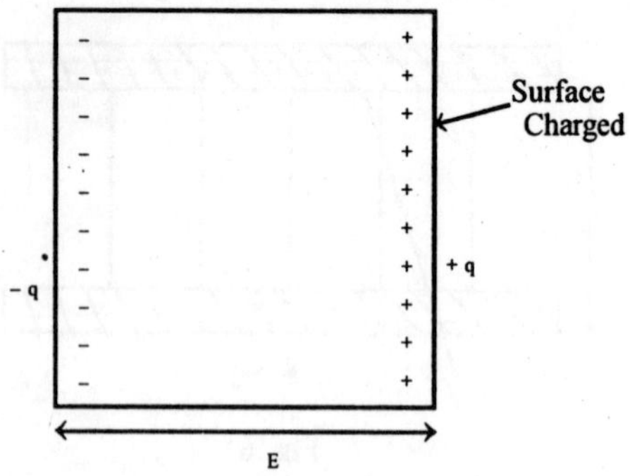

Fig. 7

This separation leads to an electric field E, which opposes the applied field E. Thus a dielectric tends to reduce the original electric field to a new value E '.

For a parallel plate capacitor

$\frac{E}{E'} = k$ Where k is known as the dielectric constant of the material.

Consider a dielectric completely filling a capacitor as shown in Fig. 8. For the gaussion surface shown

$$\epsilon_0 \oint \overline{E}.d\overline{s} = q - q' \qquad ...(1)$$

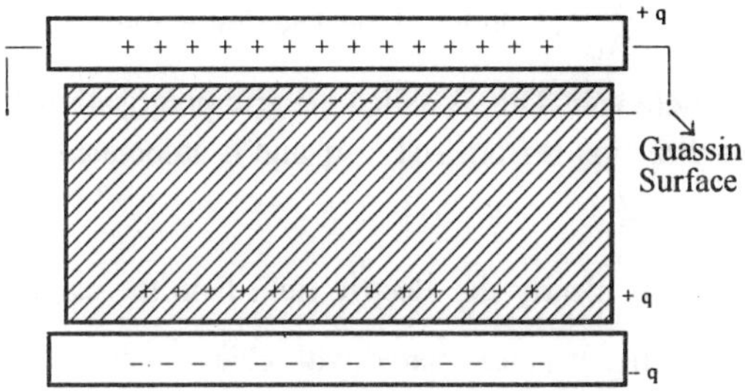

Fig. 8

Where q' is the induced surface charge on the dielectric and therefore net charge within the gaussian surface is $(q - q')$

$$\text{or } \epsilon_0 E'A = q - q' \qquad ...(2)$$

$$\text{or } E' = \frac{q}{\epsilon_0 A} - \frac{q'}{\epsilon_0 A}$$

Where dielectric is not present $E = \frac{q}{\epsilon_0 A}$

and also $\frac{E}{E'} = k$

Therefore, equation (2) becomes

$$E' = \frac{q}{k\,\epsilon_0 A} = \frac{q}{\epsilon_0 A} - \frac{q'}{\epsilon_0 A} \qquad ...(3)$$

$$q' = q\left(1 - \frac{1}{k}\right) \qquad ...(4)$$

Substituting eq. (4) in eq. (1)

$$\epsilon_0 \oint k\overline{E}.d\overline{S} = q$$

This is the modified gauss's law for dielectric medium.

ELECTRIC POLARISATION AND ELECTRIC DISPLACEMENT VECTORS

Equation (3) can be rewritten as

$$\frac{q}{A} = \epsilon_0 \left(\frac{q}{k \epsilon_0 A} \right) + \frac{q'}{A} \qquad \text{...(1)}$$

1. q'/A is clearly induced surface charge per unit area. It can be written as :

   $P = \frac{q'}{A} = \frac{q'd}{Ad}$ i.e. induced electric dipole moment per unit volume of the dielectric. This is called electric polarisation P. It is a vector whose direction is from $-q't_0 + q'$ and is associated with polarisation charges only.

2. $\frac{q}{k \epsilon_0 A}$ is clearly the electric field E in the capacitor containing the dielectric. It is associated with charge actually present whether free or polarised.

3. q/A is known as electric displace-ment D and is associate with free charge only and is represented also as lines of force between the free charges.

   Thus equation (4.38.3) becomes

   $\overline{D} = \epsilon_0 \overline{E} + \overline{P}$

   $\overline{D}$ and $\overline{p}$ can be expressed in terms of $\overline{E}$ as follows

   $\overline{D} = k \epsilon_0 \overline{E}$

   $\overline{P} = \epsilon_0 (k-1) \overline{E}$

The force of attraction between oppositely charged plates in a capacitor is given by

$$F = \frac{1}{2} \epsilon_r \epsilon_0 AE^2 \text{ Newtons}$$

CAPACITORS IN SERIES AND IN PARALLEL

The equivalent capacitance of capacitors in series is given by

$$\frac{1}{C} = \frac{1}{C_1} + \frac{1}{C_2} + \frac{1}{C_3} + \ldots$$

The equivalent capacitance of parallel system is given by

$$C = C_1 + C_2 + C_3 + \ldots$$

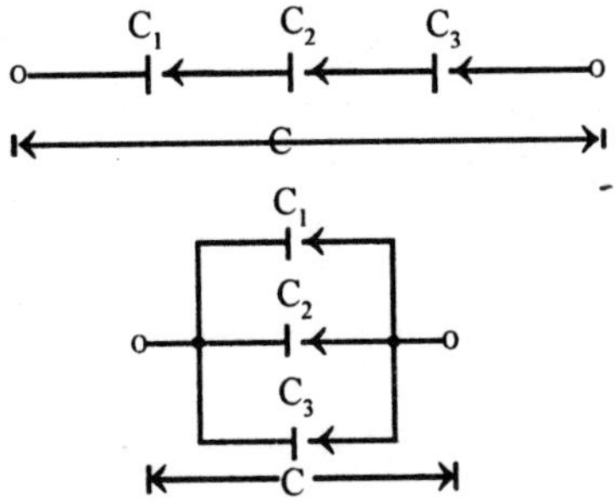

Fig. 9

CAPACITANCE OF VARIOUS SYSTEMS

Capacitance of a capacitor depends upon the ometry and medium inside it-few examples are shown in Fig 10. Capacitance of an isolated sphere of radius r in a medium of relative permittivity ϵ_r is

$$= 4\,\pi \epsilon_0 \epsilon_r \,.\, r \text{ farad}$$

Capacitance of a spherical shell with outer sphere earthed as shown in Fig. 10

$$= 4\,\pi \epsilon_0 \epsilon_r \cdot \frac{ab}{b-a} \text{ Farad}$$

Spherical capacitor with inner sphere earthed as shown in Fig. 10

$$= 4\,\pi \epsilon_0 \epsilon_r \cdot \frac{ab}{b-a} + 4\pi \epsilon_0\, b$$

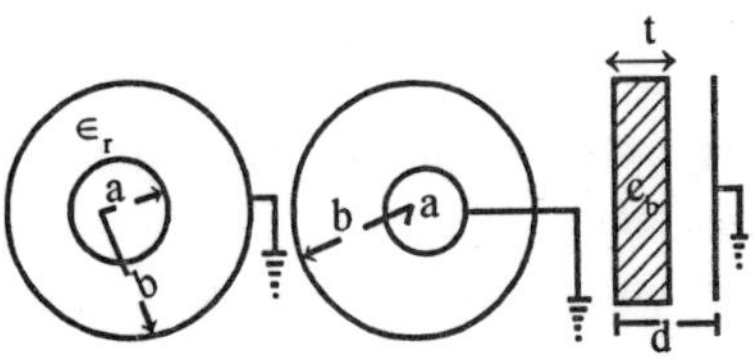

(a) (b) (c)

Fig. 10

Capacitor of uniform parallel plates with medium partly air in Fig.10.

$$= \frac{\epsilon_0\, A}{\left(d - t - \dfrac{t}{\epsilon_r}\right)} \text{ Farad}$$

ENERGY STORED IN A CAPACITOR

Work must be done to separate two equal and opposite charges. This energy is stored in the system and can be recovered if the charges are allowed to come together again. Similarly a charged capacitor has stored in it an electrical potential energy U equal to the work W required to change it.

This energy can be recovered if the capacitor is allowed to discharge. We can visualize the work of charging by imagining that an external agent pulls electrons from the positive plate and pushes them into the negative plates, thus bringing about the charge separation. Normally, the work of charging is done by a battery, at the expense of its store of chemical energy.

Suppose that at a time t, a charge q'(r) has been transferred from one plate to another. The potential difference V(t) between the plates at that moment will be q'(t)/C. If extra increment of charge dq' is transferred, the small amount of additional work needed will be

$$dw = vdq\left(\frac{q'}{C}\right)dq'$$

If this process is continued until a total charge q has been transferred, the total work will be found them.

$$W = \int dw = \int_o^q \frac{q'}{C} dq' = \frac{1}{2}\frac{q^2}{C}$$

From this equation q = CV

$$W\,(=U) = \frac{1}{2}\,CV^2$$

It is reasonable to suppose that energy stored in a capacitor resides in the electric field.

In a parallel plate capacitor neglecting fringing, the electric field has the same value for all points between the plates. Thus the energy density μ which is the stored energy per unit volume should also be uniform. It is given by

$$\mu = \frac{U}{Ad} = \frac{\frac{1}{2}CV^2}{Ad}$$

Where Ad is the volume between the plates

$$\mu = \frac{k\,\epsilon_0}{2}\left(\frac{v}{d}\right)^2 \left(\because C = \frac{k\,\epsilon_0\,A}{d}\right)$$

However v/d is the electric field strength E therefore

$$\mu = \frac{1}{2} k \,.\, \epsilon_0 \,.\, E^2$$

Although this equation is derived for the special case of parallel plate capacitor, it is true in general. If a electric field E exists at any point in space, we can think of that point as the site of stored energy in amount per unit of volume of $1/2k.\epsilon_0 E^2$.

FARADAY'S LAW IN INTEGRAL FORM

The electric field produced by the time varying magnetic field is such that the work done by it around a closed path C per unit charge in the limit that the charge tends to zero. That is, its circulation around a closed path C is equal to the negative of the time rate of change of a magnitude of flux ϕ enclosed by path C. In equation form,

$$\text{Circulation of E around } C = \frac{-d\phi}{dt}$$

The magnetic flux enclosed by C is given by the surface integral of the magnetic flux density evaluated over the surface bounded by the contour C. Thus faraday's law can be defined as

$$\oint_C \overline{E}.d\overline{l} = \frac{-d}{dt} \int_S \overline{B}.d\overline{s}$$

Faraday's law in differential form is given by

$$\nabla \times \overline{E} = \frac{\partial \overline{B}}{\partial t}$$

A magnetic field changing with time results in from of current in a loop of wire placed in a magnetic field when the magnetic field does not change in time, there is no current flow in the wire. This means that a time varying magnetic field exerts electric type forces on charges and thus produces an electric field.

INDUCTANCE

From faraday's Law

$$L = -\frac{E}{di/dt}$$

The unit of inductance is volt-Sec./Amp. which is equal to henry.

For a close pocket coil with no iron, the inductance is given by

$$L = \frac{N\phi B}{i} = \mu_0 n^2 l A \text{ where}$$

N = total number of tuns

l = length of solenoid

n = number of turns per unit length

A = Cross sectional area

INDUCTANCE IN SERIES

When the coils are connected in series as shown in Fig. 11 such that their fluxes are additive then the equivalent inductance is given by

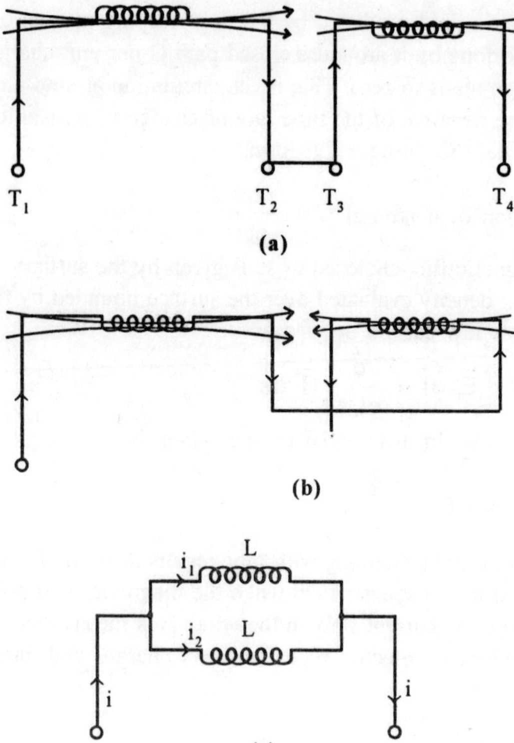

(a)

(b)

(c)

Fig. 11

$L = L_1 + L_2 + 2\,M$, where M is coefficient of mutual inductance when the fluxes oppose each other, the equivalent inductance will be

$$L = L_1 + L_2 - 2M$$

INDUCTANCE IN PARALLEL

The equivalent inductance when mutual flux helps the individual flux is given by

$$L = \frac{L_1 L_2 - M^2}{L_1 + L_2 - 2M}$$

When mutual flux opposes the individual fluxes, the equivalent inductance is

$$L = \frac{L_1 L_2 - M^2}{L_1 + L_2 + 2M}$$

MAGNETIC ENERGY STORED IN AN INDUCTANCE

By applying KVL to the circuit shown in Fig. 12.

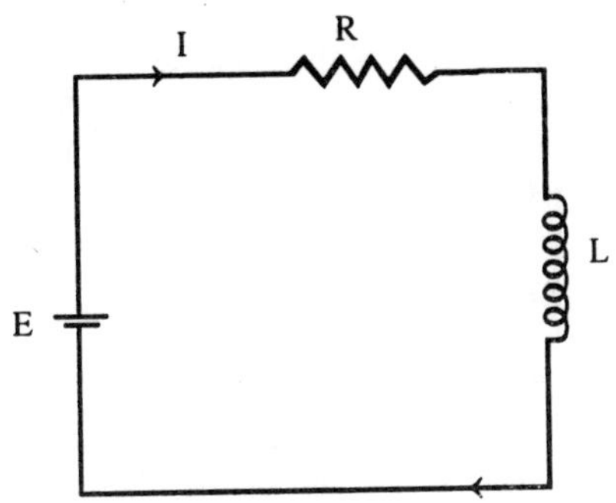

Fig. 12

$$E = iR + L\frac{di}{dt}$$

Multiply each side by i

$$Ei = i^2 R + \frac{Ldi}{di} i$$

Since $i = \frac{dq}{dt}$, the above equation implies that if a charge dq passes through the emf source in time dt, the work done on it is Edq and the rate of doing work is Edq/dt or E i. In other words, it is the rate at which the emf. Source delivers energy to the circuit, the Second term i^2R represents the rate at which energy appears as heat in the resistance R.

Energy which is not converted into heat is stored in the magnetic field of the inductor L. Thus Edi / dt is the rate dU_B/dt at which energy is stored in the magnetic field that is

$$\frac{dU_B}{dt} = L_i \frac{di}{dt}$$

or $dU_B = L\,di$

or $$U_B = \int_0^i Li\,di = \frac{1}{2} Li^2$$

Thus in an electric circuit, energy is required to maintain the flow of current but in magnetic circuit no energy is required to maintain the magnetic flux/energy is only required in creating the magnetic flux and energy supplied to create a flux is stored in magnetic field.

A.C. CIRCUITS

In a.c. circuits, the relationship $I = \frac{V}{R}$ does not hold good the variations in current and applied voltage setup magnetic and electrostatic effects respective. The effects must be taken into account in those circuits wish comparatively law voltage, heavy current circuits, magnetic effects may be very large but electrostatic effects are negligible with high voltage circuits, electrostatic effects are of appreciable magnitude. This may lead to variations in resistance thus defying $I = V/R$ relationship.

Purely Resistive Circuits

When the current flowing through a pure resistance no back emf is setup and the applied voltage has to over come the ohmic drop if iR only. The instantaneous current is

$$i = \frac{V_m \sin \omega t}{R} = I_m \sin \omega t$$

The applied voltage and current are in phase with each other.

PERIODIC TIME AND FREQUENCY

The time taken in seconds by an alternating quantity to complete one cycle is known as periodic time (T).

The number of cycles per second by an alternating quantity is known as frequency.

$$f = \frac{1}{T} \text{ cycles/seconds or Hertz}$$

ELECTRIC FIELD INTENSITY & POTENTIAL

From vector algebra it is seen that if the curl of vector is zero, the vector can be expressed as gradient of a scalar function at any point is normal to the surface passing through that point & on which the scalar function is of constant value, this implied that the vector $\overline{E}$ can be expressed as the gradient of potential i.e. V

The potential difference b/w point A & B is

$$-\Delta V = \int_A^B \overline{E}.d\overline{l}$$

$$= - E_A \,\Delta l.\cos\theta = - E_A \Delta_n$$

or $E_A = \dfrac{\Delta V}{\Delta n}$

and $\overline{E}_A = \dfrac{\Delta V}{\Delta n}\bar{\iota}_n$ (where Δn is normal vector at right angles at point A)

This means that as one travels through an electric field the rate of change of potential with charged sign is the component of $\overline{E}$ is that direction of movement or

$$E_t = \frac{-dy}{dl}$$

If $\Delta n \to 0 \left(\dfrac{\Delta V}{\Delta n}\right)\bar{\iota}_n$ by definition becomes equal to ΔV

Therefore, $\overline{E} = -\Delta V$

The Maxwell's divergence equation is

$$\Delta.\overline{E} = \frac{\rho}{\varepsilon_0}$$

Therefore $\Delta.(-\Delta V) = \dfrac{\rho}{\varepsilon_0}$

$$\Delta^2 V = \frac{\rho}{\varepsilon_0}$$

This is called poissons equation and relaties the potential at a point to the volume charge density ρ at that point. If the volume charge density is zero in a region, then.

$\Delta^2 V = 0$

This is known as Laplace's equation.

MAGNETIC DEFLECTING FORCE

If the positive test charge q_0 is fired with velocity ρ through a point p in a magnetic field $\overline{B}$.

Then the $\overline{F}$ acting on the moving charge is given by

$$\overline{F} = q_0\overline{V} \times \overline{B}$$

or $F = q_0.VB \sin\theta$

Where angle θ is the angle between $\overline{V} \times \overline{B}$. The force experienced by the moving charge duc to the magnetic field is normal to both $\overline{V} \times \overline{B}$. Therefore, there is no acceleration along the direction of motion & the magnetic field can only change the direction.

LORENTZ RELATION

When a charged particle q_0 is subjected to an electric field $\overline{E}$ & magnetic field $\overline{B}$, the resultant force acting on it is given by

$$\overline{F} = q_0\overline{E} + q_0\overline{V} \times \overline{B}$$

For a continuous charge distribution of density ρ moving with a velocity r, the force per unit volume can be defined as

$$\overline{F} = \rho\overline{E} + \overline{J} \times \overline{B}$$

Where $\overline{J}$ = volume current density and is given by $\overline{J} = \rho\overline{\imath}$.

MAGNETIC FLUX DENSITY

$\overline{F_1}$ & $\overline{F_2}$ are two non-zero force for two velocities r_1 & r_2 in different directions in the magnetic field $\overline{B}$

$$\overline{F_1} \times \overline{F_2} = \left(qr_1 \times \overline{B}\right) \times \left(q\overline{r}_2 \times \overline{B}\right)$$

$$= q_2\left(\overline{V_1} \times \overline{B_1}\overline{B}\right) r_2 - \left(\overline{r_1} \times \overline{B_1}\overline{r_2}\right)\overline{B}$$

$$= -q\left(\overline{F_1}.r_2\right)\overline{B}$$

$$\overline{B} = \frac{\overline{F_2} \times \overline{F_1}}{q\left(\overline{F_1}.\overline{r_2}\right)}$$

If the maximum force on q_0 is $\overline{F}_m$ and the direction of $\overline{r}$ is $\overline{\imath}_m$ then

$$\overline{B} = \frac{\overline{F}_m \times i_m}{q\overline{r}}$$

Therefore unit of $\overline{B} = \dfrac{\text{Newton per coulomb}}{\text{meter per second}}$

$$= \frac{\text{Newton second}}{\text{coulomb. meter}}$$

$$= \frac{\text{Newton. meter}}{\text{coulomb}} \times \frac{\text{second}}{(\text{meter})^2}$$

Now since Newton meter per coulomb is a volt, unit of B = volts – second per square meter, commonly known as weber per meter^2 or WB/m^2.

FORCE ON A CONDUCTOR

Force on a current carrying conductor in magnetic field is given by

$$\overline{F} = I\,\overline{l} \times \overline{B}$$

$$= BIl \sin \theta$$

The direction of force can be determined by applying flaming's Left Hand Rule. The thumb, forefinger and middle finger of the left hand are stretched in such away that they are at right angles to each other mutually and the prefinger points towards the direction of the magnetic field, middle finger towards the direction of the magnetic field, middle finger towards the direction of the flow of current, then thumb will be in the direction of force acting on the conductor.

Force on a Current Carrying Line

The total force experienced by a filamentary wire carrying current I as shown in Fig. 13 is given by

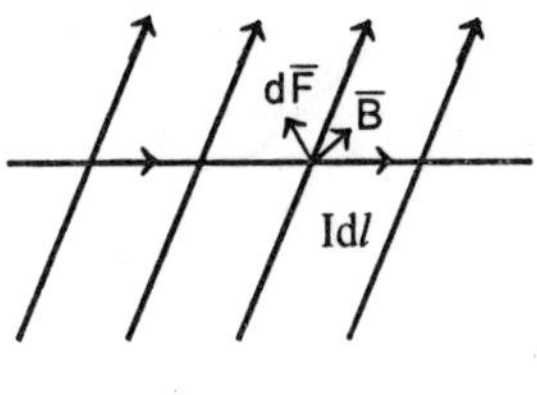

Fig. 13

$$\overline{F} = \int_{\text{wire}} \left(Id\overline{l} \times \overline{B}\right)$$

$$= = I \int_{\text{wire}} \left(d\overline{l} \times \overline{B}\right)$$

Magnetic Dipole

Consider a rectangular loop of length a and width b placed at an angle θ with uniform magnetic field $\overline{B}$. The forces $\overline{F}_2$, $\overline{F}_4$ on sides 2 and 4 equal and opposite and therefore cancel each other.

The forces $\overline{F}_1$ and $\overline{F}_3$ on sides 1 and 3 are also equal and opposite. But since they do not have the same line of action, they produce a torque τ, on the loop and thus tends to rotate it is clockwise direction about the axis xx^1. This torque is given by

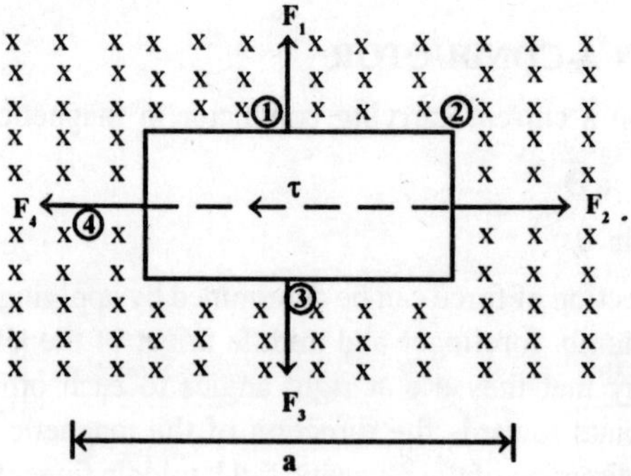

Fig. 13

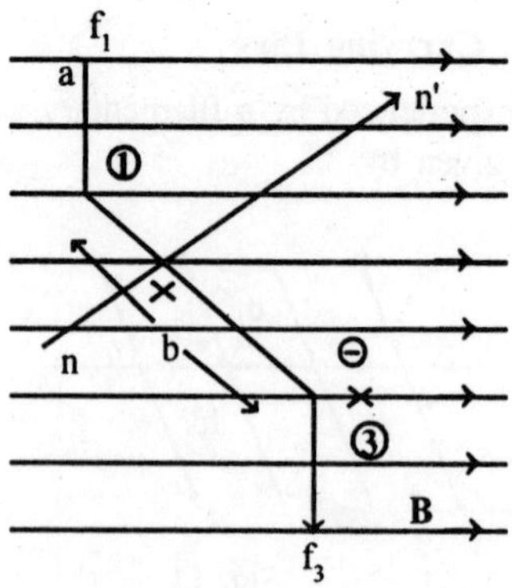

Fig. 14

τ = Force × Distance between the forces

= ia B.b sin θ

= iAB sin θ where A = Area of the loop

If there are N turns in the loop, then

$\tau = (N.i.A)\ B \sin\theta$

The terms (NiA) is called the magnetic dipole moment $\bar{\mu}$ of the loop. In vector form above equation becomes

$$\bar{\tau} = \bar{\mu} \times \bar{B}$$

The vector $\bar{\mu}$ lies along the axis of the loop. This torque acting on a current loop is basis of operation of electric motors and many electric meters.

Magnetic Potential Energy

As in the case of electric dipole described previously, work must be done to change the orientation of a magnetic diploe or current loop placed in a magnetic field. This implies that it has potential energy which is taken to be zero for $\theta = 90°$. The magnetic potential energy will be

$$U = \int_{90°}^{\theta} \tau d\theta = \int_{90°}^{\theta} \mu B \sin\theta . d\theta = -\mu B \cos\theta$$

$$U = -\bar{\mu}.\bar{B}.$$

THE BIOT–SAVART LAW

The field strength due to a small conductor of length of carrying current I at any point P as shown in Fig 15.

(i) Directly proportional to length δl.

(ii) Directly proportional to the strength of current flowing.

(iii) Inversely proportional to the square of the perpendicular distance of the point from δl.

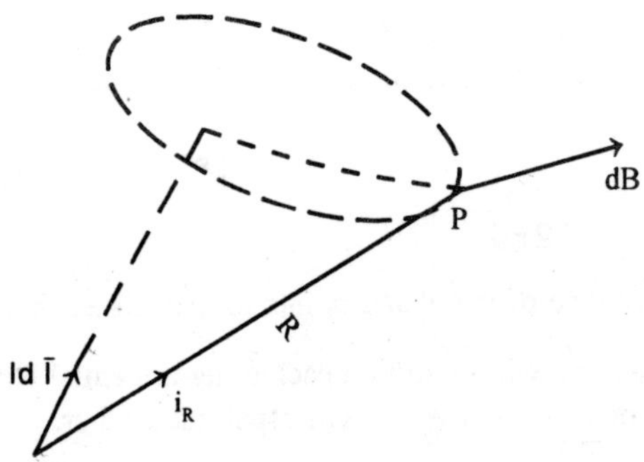

Fig. 15

Mathematically, $d\bar{b} = \frac{\mu_0 I \delta l}{4\pi r^2} i_R$

This is known as Biot-savart low is analogous to the electric field intensity of a point charge. According to this law, the magnetic flux density at a point P due to a current element is normal to the plane containing the element and the line joining the element and point. Clearly it is circular to the straight line axis along the current element.

The direction of flux density can be known by visualising right hand grabbing the filament wire and the thumb pointing in the direction of current. The direction in which the finger curl is the direction of field. Alternately it is the same as the turning the right hand screw as it advances in the direction of current.

The magnetic flux density due to a wire of any length is given by

$$\bar{B} = \frac{\mu_0}{4\pi} \int_C \frac{Id l \times I_R}{R^2}$$

Where the integral is taken along the contour C of the wire.

4.20.1 Magnetic field due to a current carrying wire

The magnetic flux density at point P at a distance r from infinity long strength wire carrying current I as shown in Fig. 16.

$$B = \frac{\mu_0 I}{2\pi r} I_\phi$$

Where I_ϕ is a unit vector in the direction of magnetic field 4.20.2 Force between two parallel conductors

Two long parallel wire separated by a distance d and carrying currents ia and i_b, a length l of wire b will experience a side ways magnetic force whose magnitude is

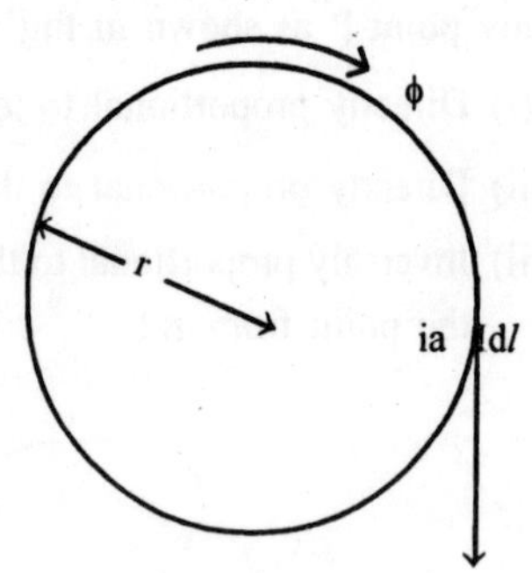

Fig. 16

$$\bar{F} = i_b l \bar{B}_a = \frac{\mu_0 I_a l_b}{2\pi d}$$

$\bar{F}$ lies is the plane of wires and points to the left as shown in Fig. 17.

The forces that two wires exert on each other are equal and opposite for anti-parallel currents, the two wires will repel each other.

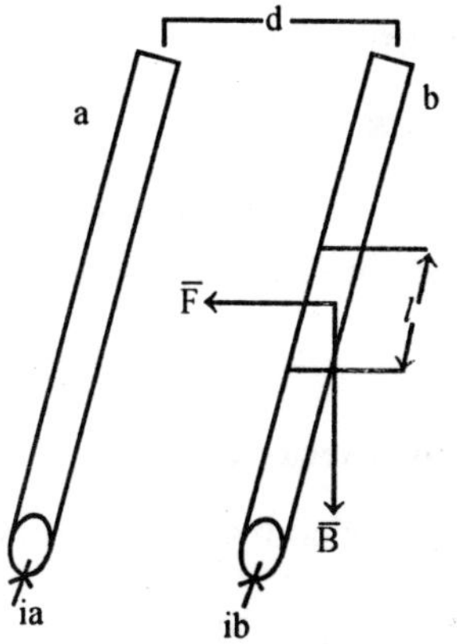

Fig. 17

AMPERE'S LAW

Consider a circular path C of radius r in the plane normal to the wire and centered at the wire as shown in fig 4.20.1. Take an infinitesimal element of length dl on this path. then

$$\overline{B}\,d\bar{l} = \left(\frac{\mu_0 I}{2\pi r}\bar{I}_\phi\right) dl\; i\,\phi \quad = \frac{\mu_0\, I\, dl}{2\pi\, r}$$

Therefore, $\oint_c \overline{B}\,d\bar{l} \oint_c \dfrac{\mu_0 I d\bar{l}}{2\pi r} = \dfrac{\mu_0 I}{2\pi r}\, 2\pi r = \mu_0 I$

When generalised over any arbitrary closed path ground the wire, this equation is known as ampere's law. It gives information about the current enclosed by the path C. If the arbitrary path does not enclose current then

$$\oint_c \overline{B}\,dl = 0$$

4.21.1The ampere's law in differential form

The current enclosed by arbitrary closed path C is given by the surface integral of

the current density over any surface S bounded by closed path C. That is

$$I = \int_s \overline{J}.d\bar{s}$$

According to Ampere's law

$$\oint_c \overline{B}.d\bar{l} = \mu_0 \int_s J.d\bar{s}$$

If the path C is shrunk to a very small size ΔC so that the surface are bounded by it becomes very small

$$\oint_{\Delta C} \overline{B}.d\overline{l} = \mu_0 \int_{\Delta S} \overline{J}.d\overline{s}$$

Since surface area ΔS is very small, current density can be considered to be uniform over the surface so that

$$\oint_{\Delta s} \overline{J}.d\overline{s} \cong \overline{J}.\overline{I}_n \Delta S$$

Where in is the normal vector to ΔS

Dividing the above equation by ΔS and letting $\Delta S \to 0$

$$\lim_{\Delta S \to 0} \frac{\oint_{\Delta C} \overline{B}.d\overline{l}}{\Delta S} = \lim_{\Delta S \to 0} \frac{\oint_{\Delta S} \overline{J}.d\overline{s}}{\Delta S}$$

$$= \mu_0 \lim_{\Delta S \to 0} \frac{\overline{J}.\overline{I}n\, \Delta S}{\Delta S}$$

$$= \mu_0 \overline{J}.\overline{I}_n$$

Now, $\left|\nabla X \overline{B}\right|$ = maximum value of $\lim_{\Delta S \to 0}$

$$= \frac{\oint_{\Delta C} \overline{B}.d\overline{l}}{\Delta S} = \mu_0 \left|\overline{J}\right|$$ and Direction of $\nabla X \overline{B}$ = direction of $\overline{J}$

so that $\nabla X \overline{B} = \mu_0 \overline{J}$

MAGNETIC FIELD DUE TO CURRENT CARRYING CONDUCTOR

When current flows through a conductor, magnetic field is produced around it. If the direction of current is upwards as shown by a dot in Fig. 18, then the direction of lines of force is anti clockwise.

On the other hand, if the current is flowing down words (shown by a cross), the direction of force is clockwise.

The direction of flow of current, is rotated in such a direction that the hand of the screw mores in direction of flow of current, then the direction of rotation of the lead will be the direction of magnetic field.

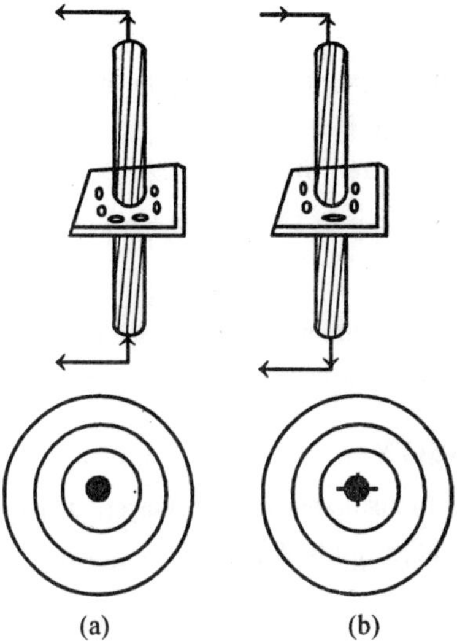

Fig. 18

THE GAUSS LAW FOR MAGNETISM

According to this law, the flux ϕ_B due to the magnetic field B through any closed gaussion surface must be zero. That is

$$\phi_B = \oint \bar{B}.d\bar{s} = 0$$

THE OHM'S LAW

When an electric field is applied across a conductor it does not finally accelerate the electrons in the conductor because they keep on colliding with the static ions in the conductor as they pass through the array of ions. During these collisions. The kinetic energy of the accelerating electrons is transferred into vibrational energy of the ions which start vibrating about their mean position.

Thus, the electrons acquire only a constant average drift speed in the direction opposite to the applied electric field. This flow of electrons i.e. charge constitute electric current. Inside the conductor, current is characterised by current density j & it is a vector pointing in the direction of current vector.

The magnitude of the current density for all points in the conductor in which the current is uniformly distributed across the cross - sectional area A will be

$$J = \frac{I}{A}$$

The vector J at any point is oriented in the direction in which a positive charge would more at that joint.

More generally, current its the flux of vector j over a particular surface in a conductor i.e.

$$\bar{I} = \int \bar{j}.d\bar{s}$$

Where $d\bar{s}$ is a small element of surface area on the surface & the integral is performed over the whole surface under consideration.

CURRENT DENSITY

The general relationship between the current density vector J in a conductor and the electric field $\bar{E}$ applied across it is given by ohm's law in the form.

$\bar{J} = \sigma\bar{E}$ where σ is the electrical conductivity of the material.

Since the applied electric field corresponds to a potential difference across the conductor, the ohm's law alternatively states that the ratio of potential difference V, between two points on a conductor to the current I flowing between there points is a constant, provided the temperature of the conductor remains constant. This law hold if voltage current characterised of the conductor is linear. The constant is known resistance R of the material or

$$\frac{V}{I} = R$$

Note that the ohm's law is not uniformly true.

RESISTIVITY

Resistivity ρ is a characteristic of a material & is defined as ratio of electric field applied across a conductor to the current density inside the conductor i.e.

$$\rho = \frac{E}{J}$$

Now consider a cylindrical conductor of length l & cross - sectional area A. If a potential difference V is applied across its ends, then

$$i = \int \bar{j}.d\bar{s} = jA$$

and $V = -\int_a^b \bar{E}.d\bar{l} = -El$

Therefore $R = \frac{V}{I} = \frac{El}{JA}$

$$\rho = \frac{l}{A}$$

Typical value for copper and aluminium are 1.7×10^{-8} and $2\text{-}8 \times 10^{-8}$ respectively. In case of conductors when temperature rises the thermal energy given to a material increases its vibrational energy of the ions which starts vibrating at a higher rate about their means positions.

This leads to a decrease in the mean free path available for the free electrons - thus these electrons find it difficult to more along the applied electric field, thus leading to decrease in current density and hence increased resistivity of the material. This phenomenon resistance for a unit charge in temperature i.e.

$$\theta = \frac{d\rho}{dT} \times 100$$

Conductors have a positive temperature coefficient (PTCs) while semi conductors have negative temperature coefficient (NTCs)

KIRCHOFF'S CURRENT LAW (KCL)

In any electrical network, the algebric sum of the current meeting at a point is zero.

$$\Sigma i = 0$$

KIRCHOFF'S VOLTAGE LAW (KVL)

The algebric sum of the products of current and resistance in each of the conductors in any closed mesh in network plus the algebric sum of the emfs in that path is zero.

$$\Sigma IR + \Sigma_{emf} = 0$$

EQUIVALENT RESISTANCE

The equivalent resistance R of n number of resistance R_1, R_2, R_n connected in series is

$$R = R_1 + R_2 + ...R_n$$

The equivalent resistance R of n number of resistances R_1, R_2, R_n connected in parallel is given by

$$\frac{1}{R} = \frac{1}{R} + \frac{1}{R_2} + \ldots + \frac{1}{R_1}$$

Transfer of Energy in Electric Circuit

Consider a current i flowing in a device due to a potential V across its terminals. Let a charge dq moves through the device. This will lead to a decrease in potential energy of the charge by dq. V. This change in potential energy will transfer to some other form of energy depending upon the type of device used. The amount of energy transfer in time dt will be

$$du = dq \cdot V$$

The rate of energy transfer will be

$$P = \frac{du}{dt} = \frac{dq}{dt} \cdot V = V \cdot i$$

If the device is a battery, the energy will be stored as chemical energy, If it is an electric motor the energy will be converted into mechanical work done. If the device is a resistor, the electrons in the resistor will gain this energy but lose as heat, due to collisions with the ions. Thus the heat dissipated when a current i flows in resistor R is

$$P = IV = I^2R$$

Now volt = Joule/Coulomb

$$\text{Therefore, } P = iV = \frac{\text{Coloumb}}{\text{Sec}} \; \frac{\text{Joule}}{\text{Coloumb}}$$

or Volt - Ampere = Joule/Sec.

PRIMARY CELLS AND SECONDARY CELLS

An electric cell is a device in which an electric difference of potential is established between two electrodes as a result of chemical reaction between electrodes and electrolyte. Simply unit is known as a cell. A combination of several units connected in series electrically is known as a battery. Cells are of two types primary cells, and secondary cells.

PRIMARY CELLS

It consists of a vessel containing a liquid called electrolyte and two dissimilar metals plates known as electrodes immersed in the liquid. As a result, a potential difference is setup between the two plates due to the chemical reaction. A simplified sketch is shown in Fig. 19.

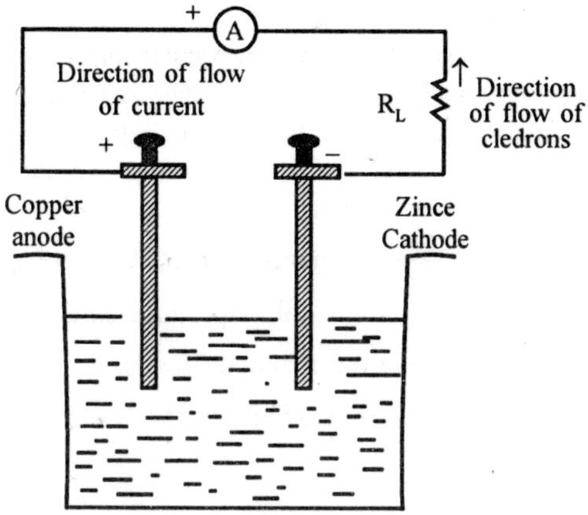

Fig. 19

THERE ARE TWO MAIN DEFECTS IN THE TYPE OF CELLS LOCAL ACTION AND POLARISATION

Local Action

Commercial zinc used as electrode contains various impurities like carbon, lead etc. which leads to the formation of countries little local cells with the zinc plate, giving rise to a local currents. Due to these local currents, zinc is wasted even when the cell is not delivering any current to the external circuit.

The defect can be avoided by amalgamating the zinc plate by which a thin layer of mercury amalgam deposited over the impure zinc does permit this impurities to come to the surface.

Polarisation

During the chemical action in the cell, hydrogen gas is liberated and few of the hydrogen bubbles get deposited on the copper plate with the result that it gets gradually covered with hydrogen gas layer. Due to this, the cell is weakened. This phenomenon can be avoided either by corrugating the positive plate or agitating of the liquid or of the positive plate so that accumulation of hydrogen an positive plate is avoided. Using a depolarising agent or substituting the hydrogen by some other substance which does not give counter emf are other methods to minimise polarisation.

The characteristics of a good cells are high and constant voltage small internal resistance constant current & it should be cheap and does not give corrosive fumes.

A.C. CIRCUIT

The voltage changes its polarity at regular intervals of time is called an alternating voltage. The graph of instantaneous voltage and time is called waveform. Alternating voltage may be in square waveform, triangular waveform, rectangular waveform etc. However, we choose sine waveform because of the following reasons;

(i) It produces least disturbance in electric circuits.

(ii) It is easier and economical to produce sine waveforms.

(iii) It can produce rotating magnetic fields.

(iv) It can be easily represented in phasor form and easier for mathematical calculations.

GENERATION OF ALTERNATING VOLTAGE AND CURRENT

When a conductor is rotated in magnetic field an alternating e.m.f. is generated in the conductor. An alternating e.m.f. will be generated if magnetic field is changing and coil is stationary. The generated e.m.f. will depend upon strength of magnetic field, number of turns in the coil and speed at which the coil or magnetic field rotates.

Let us consider that, the coil has N number of turns and is rotating in uniform magnetic field at the angular velocity of ωrad/sec. Let initial position of coil is an X-X axis. After the time of *t* seconds, it has achieved the position of 1-1 after rotating by angle θ.

Thus $\theta = \omega t$

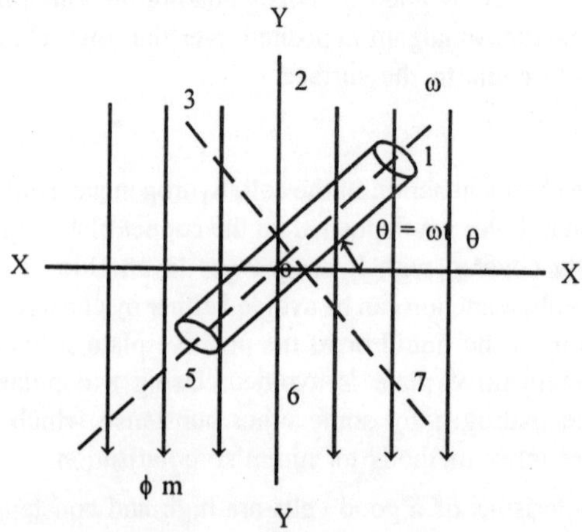

Fig. 20

At the instant 1-1, flux perpendicular to the coil is = $\phi_m \cos \omega t$ where ϕ_m is the maximum possible flux. But Faraday Law states that

$$e \text{ (generated e.m.f.)} = -N\frac{d\phi}{dt}$$

$$= -N\frac{d}{dt}(\phi_m \cos \omega t)$$

$$e = N\,\phi_m\,\omega t \sin \omega t$$

$$e = E_{max} \sin \omega t$$

where

$$E_{max} \text{ (max. value of e.m.f. generated)} = N\phi_m\omega$$

$$= \omega\,N\,B_m\,A$$

$$= 2\pi f\,N\,B_m\,A$$

where

B_m = Maximum flue density (ω_b/m^2)

A = Area of coil (m^2)

f = frequency of rotation

The graph of e.m.f. induced w.r.t. time is sinusoidal as shown in figure 2.1. It is clear that when the coil is at position 0 and 4, the generated e.m.f. is zero. However at position 2, it is maximum positive and at position 6, generated e.m.f. is maximum negative.

This alternating e.m.f. can also be represented by

$$e = E_{max} \sin \omega t$$

$$= E_{max} \sin \theta \text{ (as } \theta = \omega t)$$

$$= E_{max} \sin (2\pi f t) \text{ (as } \omega = 2\pi f)$$

$$e = E_{max} \sin \left(\frac{2\pi}{T}\right)t \qquad \text{(as f = 1/T)}$$

Similarly alternating current is represented by

$$i = I_{max} \sin \left(\frac{2\pi}{T}\right)t$$

R.M.S. (ROOT MEAN SQUARE VALUE)

R.M.S. value of an alternating quantity is equal to 0.707 times the maximum value

$$I_{r.m.s.} = \frac{I_m}{\sqrt{2}} = 0.707 \times I_m$$

$$\& \qquad V_{r.m.s.} = \frac{V_m}{\sqrt{2}} = 0.707 \times V_m$$

AVERAGE VALUE

As average value for symmetrical waveform for a complete cycle is zero.

Hence for sinusoidal wave it is taken over half cycle only. The average value for a sinusoidal quantity is 0.637 times the maximum value.

$$I_{av} \Rightarrow 0.637 \times I_m$$

$$\& \qquad V_{av} = 0.637 \times V_m$$

FORM FACTOR

$$\text{Form factor} = \frac{\text{r.m.s. value}}{\text{average value}}$$

$$\Rightarrow \frac{0.707\, I_m}{0.637\, I_m}$$

$$= 1.11 \text{ (for Sine Walle)}$$

AMPLITUDE FACTOR OR CREST FACTOR

$$\text{Amplitude factor} = \frac{\text{Maximum Value}}{\text{r.m.s. Value}}$$

$\Rightarrow$ 1.414 respectively

APPLICATION OF A.C. FOR BASIC ELEMENTS

(a) **Resistance :** When alternating voltage

$e = E_{max} \sin \omega t$ is applied to pure resistance of value R, then the current i starts flowing:

$$i = \frac{e}{R} = \frac{E_{max}}{R} \sin \omega t.$$

or $\qquad i = I_{max} \sin \omega t.$

where $I_{max} = \dfrac{E_{max}}{R}$

From the above equations, it is clean that the current and the voltage remains in same phase, and the phase angle between voltage and current is zero.

The power factor is unity.

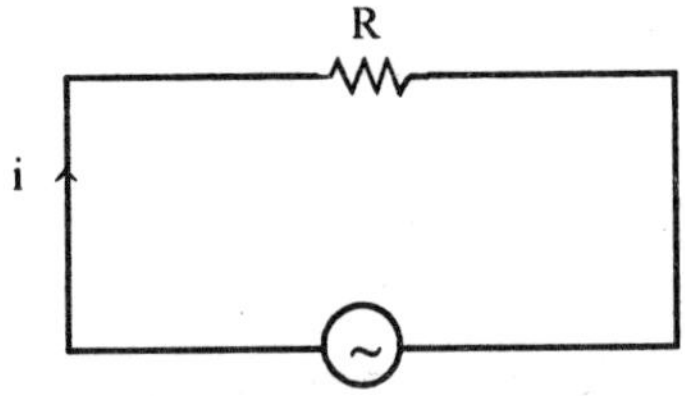

$e = E_{max} \sin \omega t$

Fig. 21

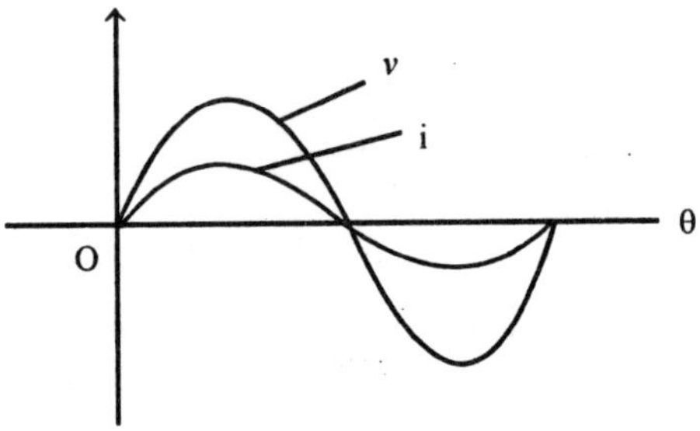

Fig. 22 : Waveform

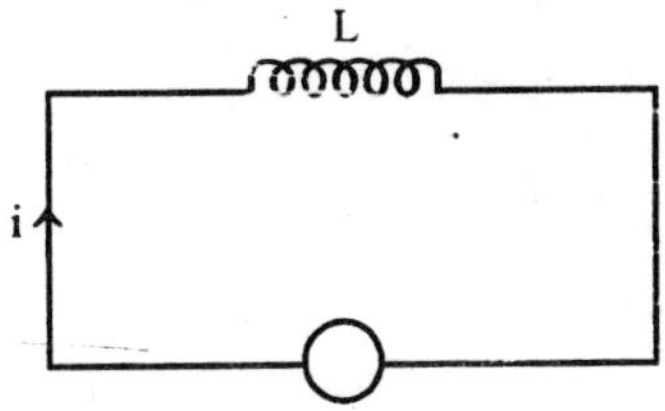

Fig. 23 : Phasor diagram

Instantaneous power $P = e \times i$

and average power $= E \times I = I^2.R$ watts

where I and E are r.m.s. values of voltage and current are in phase with each other.

(b) **A.C. Through Pure Inductance :** When alternating voltage is applied to pure inductance a back emf is produced due to self inductance.

This back emf always oppose change in current. The applied voltage has to over come this self induced emf only.

$$e = \frac{Ldi}{dt} = E_{max} \sin \omega t$$

$$di = \frac{E_{max}}{L} \sin \omega t \, dt$$

$$i = \frac{E_m}{L} \int \sin \omega t \, dt$$

$$= \frac{E_m}{\omega L} (- \cos \omega t)$$

$$= \frac{E_m}{\omega L} \sin (\omega t - \pi/2)$$

$$i = I_{max} \sin (\omega t - \pi/2)$$

$$\text{where } I_{max} = \frac{E_{max}}{\omega L}$$

This indicates that current lags behind the applied voltage by quarter cycle or phase difference between two is $\pi/2$ as shown in Fig. 24.

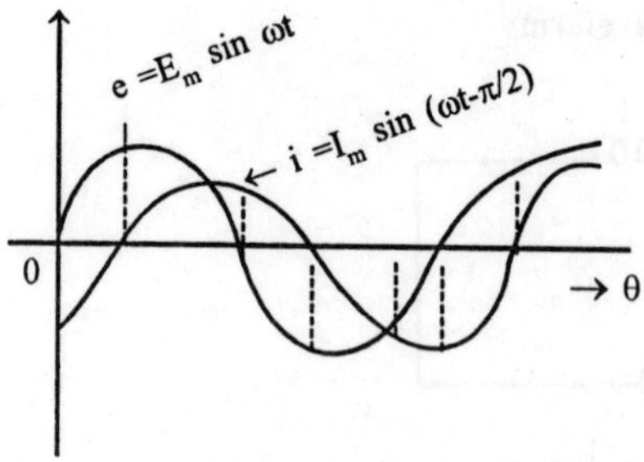

Fig. 24

Fig. 25

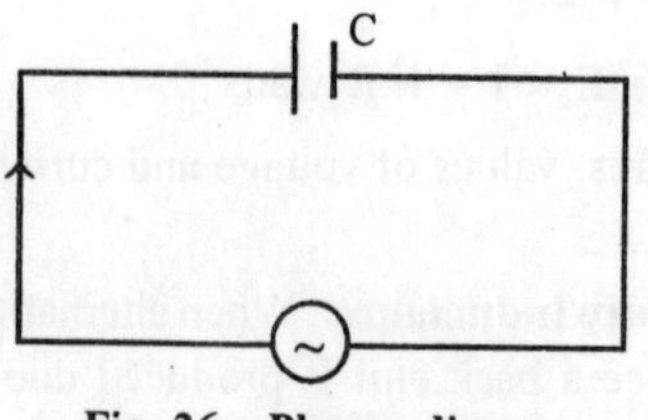

Fig. 26 : Phasor diagram

The angle of power factor = 90°

Thus power factor = cos 90° = 0

Hence power = VI cos ϕ = 0

Hence pure inductance does not consume active power. The term 'ωL' plays the part of resistance. This is called inductive reactance and is denoted by X_L. The unit of this reactance is ohms.

$$X_L = \omega L = 2\pi f L$$

(c) **A.C. Though Pure Capacitance:** In Fig. 27 alternating voltage $e = E_{max} \sin \omega t$ is applied to the capacitance of C farads. Charging current is capacitance is given by

i = C × Rate of change of potential difference

$$= C\frac{de}{dt} = C\frac{d}{dt}(E_{max} \sin \omega t)$$

$$= C\omega E_{max} \cos \omega t$$

$$= \frac{E_{max}}{1/C\omega} \sin(\omega t + \pi/2)$$

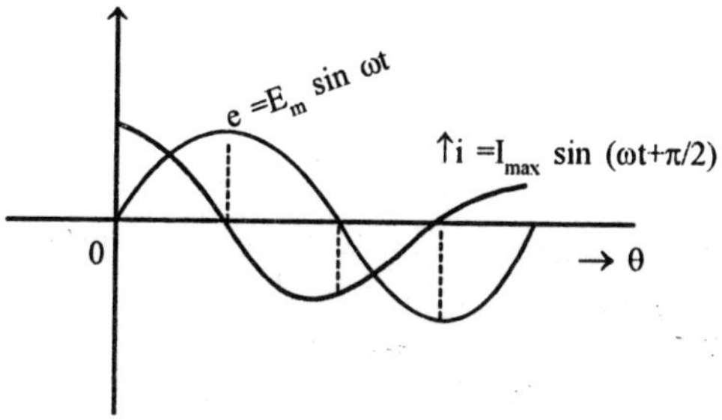

Fig. 27

$$i = I_{max} \sin(\omega t + \pi/2)$$

$$\text{where } I_{max} = \frac{E_{max}}{1/\omega C}$$

The term i/ωC is called capacitive reactance and is denoted by X_C. It is expressed in ohms.

The current in pure capacitance leads to the applied voltage by 90°.

Thus the angle of power factor = 90°

Power factor = cos 90° = 0

Thus power consumed = P = VI cos ϕ = 0

Hence pure capacitance does not consume any active power.

AC SERIES CIRCUITS

Inductance, capacitance and resistance may be connected in series depending an requirement. Thus we have

(i) R-L series circuits

(ii) R-C series circuits

(iii) R-L-C series circuits

Active power = VI cos ϕ

Reactive power = VI sin ϕ

Apparent power = VI

RESONANCE IN R-L-C CIRCUIT

A series circuit is said to be in electrical resonance when its net reactance is zero. The frequency, f_0, at which this happens is known as resonance frequency.

(i) At resonance net reactance

$$X = X_L - X_C = 0 \text{ or } X_L = X_C$$

(ii) Hence $V_L = V_C$ or $V_L - V_C = 0$

(iii) Impedance at resonance is only resistance $Z = R$

(iv) Line current is maximum and

$$I_m = \frac{V}{R}$$

(v) As $X_L = X_C$; $\omega L = \frac{1}{\omega C}$

$$\therefore \quad \omega^2 = \frac{1}{LC}$$

$$\text{or} \quad (2\pi f_0)^2 = \frac{1}{LC}$$

$$\text{or} \quad f_0 = \frac{1}{2\pi\sqrt{LC}}$$

A.C. PARALLEL CIRCUIT

In parallel circuit, two or more elements of impedances are connected across the supply.

Fig. 28, shows resistance, inductance and capacitance connected in parallel.

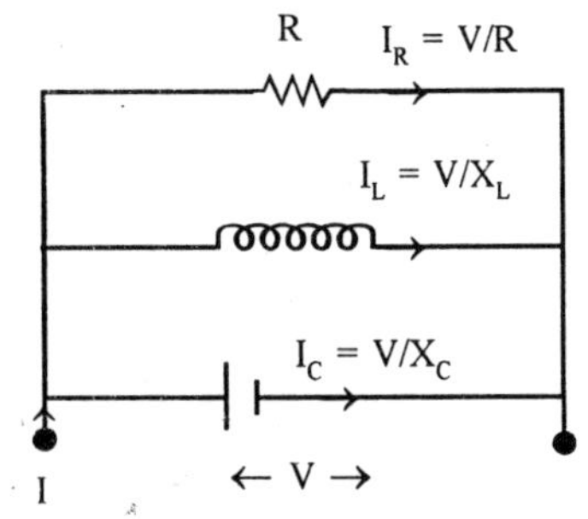

Fig. 28

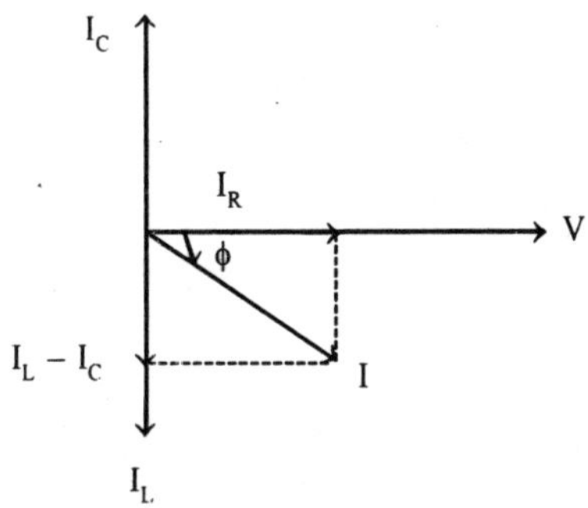

Fig. 29 : Phasor diagram

A.C. SERIES PARALLEL CIRCUIT

Fig. 30 shows that the Z_2 & Z_3 are connected in parallel and Z_1 is connected in series with the parallel combination.

$$I_1 = I_2 + I_3$$

and $$I_1 = \frac{V}{Z\ Total}$$

where Z total = total impedance of the circuit

$$= Z_1 + [Z_2 \parallel Z_3]$$

$$= Z_1 + \frac{Z_2 Z_3}{Z_2 + Z_3}$$

$$\text{current } I_2 = \frac{Z_3}{Z_2 + Z_3} \cdot I_1$$

$$\text{current } I_3 = \frac{Z_2}{Z_2 + Z_3} \cdot I_1$$

Admittance of the circuit is defined as the reciprocal of its impedance. The unit of admittance is mho.

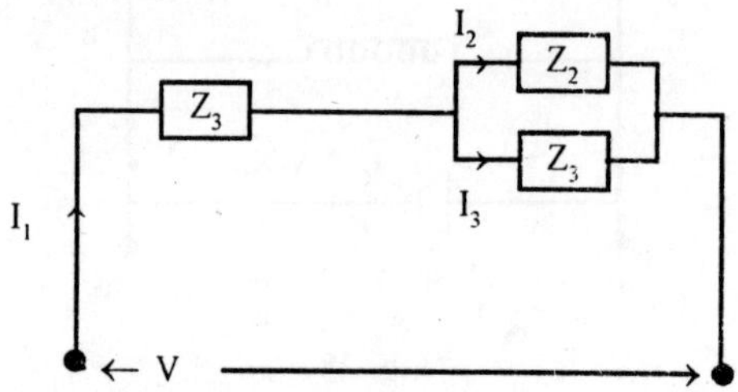

Fig. 30

GENERATION OF THREE PHASE VOLTAGE

In case of three phase alternator (AC Generator) field winding an rotaı and stator has three windings of equal turns. As shown in Fig. 31, coils aa, bb and cc are of equal number of turns and places 120° apart. When two pole rotor rotates, voltages induced in three coils are

$e_a = E \angle 0 = E_{max} \sin \omega t$

$e_b = E \angle -120^\circ = E_{max} \sin (\omega t - 120^\circ)$

$e_c = E \angle -240^\circ = E_{max} \sin (\omega t - 240^\circ)$

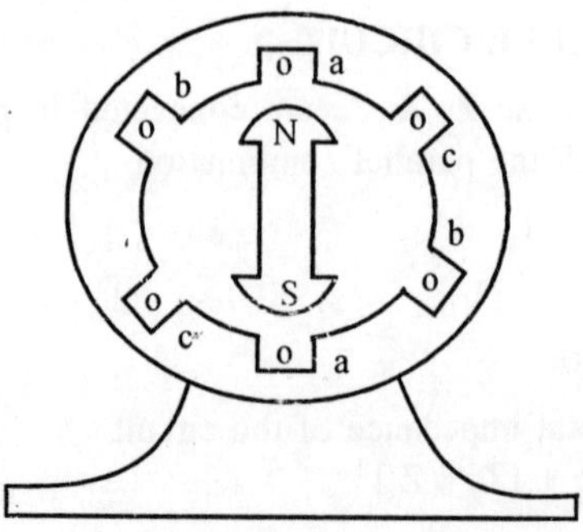

Fig. 31

The vector sum of three emf at any point is zero. These three voltages can be represented by phasor diagram as shown in Fig. 32.

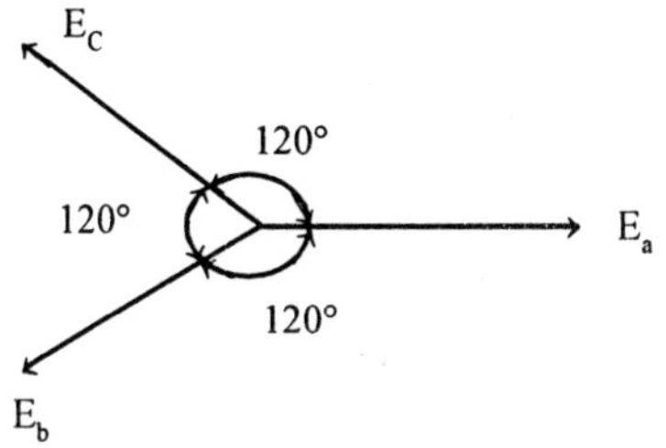

Fig. 32

(i) **Balanced System :** The magnitude of all three voltages are same i.e.

$E_a = E_b = E_c$

And they are displaced by 120° apart. This type of supply is called balanced three phase supply.

(ii) **Unbalanced Supply :** If the magnitude or phase angle is not same, then it is called unbalanced supply.

(iii) **Phase Sequence :** The order in which the phase voltages reaches their peak or maximum value is known as phase sequence. For the given voltages, we say that phase sequence is a, b, c. However the colours are used for three phases.

They are red, yellow and blue, correspondingly phase sequence is RYB.

REASONS FOR USING THREE PHASE SUPPLY

(i) In single phase supply power is pulsating, however in three phase supply power is constant.

(ii) Output of three phase machine is more than single phase for given volume and weight.

(iii) Three phase machines are smaller than single phase machines for given output.

(iv) Power transmission is cheaper for three phase machines.

(v) Three phase supply can produce rotating magnetic field which is very useful for the motors.

STAR CONNECTION

(1) **Voltages and Current Relations :** Voltages induced is each winding are called as phase voltages. Thus V_{RN}, V_{YN}, V_{BN} are phase voltages. Voltages available at any pair of terminal is called line voltage.

Thus V_{RY}, V_{YB} and V_{BR} are line voltages.

I_R, I_Y, I_B are phase and line currents. In other words

$$I_{ph} = I_L$$

From phasor diagram

$$= V_{RN} - V_{YN}$$

$$= 2\ V_{RN} \cos 30^{\circ}$$

$$= 2\ V_{ph} \times \frac{\sqrt{3}}{2}$$

$$V_L \Rightarrow \sqrt{3}\ V_{ph}$$

thus * line voltage

$$= \sqrt{3} \times \text{phase voltage}$$

* line current = phase current
* line voltages are 120° apart
* line voltages are 30° ahead of their respective phase voltages

(2) **Power :** Total power

$$= 3 \times \text{power per phase}$$

$$= 3 \times V_{ph} \times I_{ph} \cos \phi$$

$$P = \sqrt{3}\ V_L \times I_L \cos \phi$$

(3) **Neutral Point :** In star connection neutral point is always available

DELTA CONNECTION

(1) **Voltages and Current Relations :** V_{RY}, V_{YB} V_{BR} phase and line voltages. In other word phase voltages are same

$$V_{ph} = V_L$$

I_R, I_Y, I_B are line currents and I_1, I_2, I_3 are phase currents.

From phasor diagram

$$I_R = I_1 - I_3$$

$= 2\ I_1 \cos 30^\circ$

$= \sqrt{3}\ I_{ph}$

* Thus line voltage

$= \sqrt{3}$ phase current

* line voltages = phase voltages
* line currents are 120° apart
* line currents are 30° behind their respective phase currents

(2) **Power :** Total power

$= 3 \times$ power per phase

$= 3 \times V_{ph} \times I_{ph} \cos \phi$

$= \sqrt{3}\ V_L \times I_L \cos \phi$

(3) **Neutral Point :** In delta connect

SOLVED EXAMPLES

Example 1:

Determine z-transform of the signal

$$x[n] = \begin{cases} 1 & n = -1 \\ 2 & n = 0 \\ -1 & n = 1 \\ 1 & n = 2 \\ 0 & \text{otherwise} \end{cases}$$

use the z-transform to determine DTFT of x[n].

Solution:

The formula is

$$X(z) = \sum_{n=-\infty}^{\infty} x(n) z^{-n}$$

we get

$X(z) \Rightarrow x(z)\ z^{-2} + x(1)\ z^{-1} + x(0)\ z^0 + x\ (-1)\ z^1 + x(-z)\ z^2$

$\Rightarrow z + 2 - z^{-1} + z^{-2}$

We obtain the DTFT by putting $z = e^{i\Omega}$ we get

$X(e^{i\Omega}) = e^{i\Omega} + 2 - e^{-i\Omega} + e^{-i2\Omega}$

The most commonly encountered form of the z-transform in engineering applications is a ratio of two polynomials in z^{-1}, as shown by the rational function

$$X(z) = \frac{b_0 + b_1 z^{-1} \ldots + b_M z^{-M}}{1 + a_1 z^{-1} + \ldots + a_N z^{-N}}$$

It is useful to X(z) as a product of the terms involving the roots of the nominator and denominator polynomials. i.e.

$$X(z) = \frac{b_0 \pi_{k=1}^{M} \left(1 - c_k z^{-1}\right)}{\pi_{k=1}^{N} \left(1 - d_x z^{-1}\right)}$$

The c_k are the roots of the numerator polynomial and are termed the zeros of X(z). The d_x are the roots of denominator polynomials and are termed as poles of X(z). Zero locations are denoted by 'O' symbol and poles locations with the 'x' symbol in the z-plane. The pole and zero locations completely specify X(z) except for the gain factor b_0.

Problem Determine the z-transform of the signal

$x[n] = \alpha^n u[n]$

Example 2:

Depict the ROC and pole and zero location of X(z) in the z-plane.

Solution:

Substituting x[n] is

$$X(z) = \sum_{n=-\infty}^{\infty} x(n)\, z^{-n}$$

$$= \sum_{n=-\infty}^{\infty} \left(\frac{a}{z}\right)^n$$

This is a geometric series of infinite length in the ratio α/z. The sum converges provided

$|\alpha|z|<|$ or $|z|>|\alpha|$ Hence

$$X(z) = \frac{1}{1 - \alpha z^{-1}} \quad |z| > |\alpha|$$

$$= \frac{z}{z - \alpha}$$

There is a pole at z = α and a zero at z = 0 and the shaded portion in the ROC an z plane.

As with the Laplace transform, the expression for X(z) does not corresponds to a unique time signal unless the ROC is specified. This means that two different signals may have identical z-transform but two different ROC's.

Example 3:

Determine the z-transform of the signal

$y[n] = -\alpha^n u[-n-1]$

Comment on ROC and the pole zero location in z-plane.

Solution:

Substitute y[n] into the equation

$$X(z) = \sum_{n=-\infty}^{\infty} x[n]\, z^{-n}$$

$$X(z) = \sum_{n=-\infty}^{\infty} -\alpha^n\, u[-n-1]\, z^{-n}$$

$$= -\sum_{n=-\infty}^{-1} \left(\frac{\alpha}{z}\right)^n$$

$$\Rightarrow -\sum_{k=1}^{\infty} \left(\frac{z}{\alpha}\right)^k$$

$$\Rightarrow 1 - \sum_{k=1}^{\infty} \left(\frac{z}{\alpha}\right)^k$$

The sum converges for $\left|\frac{z}{\alpha}\right| < \Big|$ or $|z| < |\alpha|$

Hence $Y(z) = 1 - \dfrac{1}{1 - z\,\alpha^{-1}} \quad |z| < |\alpha|$

$$= \frac{z}{z-\alpha} \quad |z| < |\alpha|$$

Note that Y(a) is identical to X(z) even though the time signals are quite different.

Only the ROC differentiates them. We must know the ROC to determine the correct inverse z-transform.

Example 4:

Determine the z-transform of

$$x[n] = -u[-n-1] + \left(\frac{1}{2}\right)^n u[n]$$

Comment on ROC & pole-zero location of X(z) in the z-plane.

Solution:

$$X(z) = \sum_{n=-\infty}^{\infty} \left[\left(\frac{1}{2}\right)^n u[n]\, z^{-n} - u[-n-1]\, z^{-n}\right]$$

$$= \sum_{n=-\infty}^{\infty} \left(\frac{1}{2z}\right)^n - \sum_{n=-\infty}^{\infty} \left(\frac{1}{z}\right)^n$$

$$= \sum_{n=\infty}^{\infty} \left(\frac{1}{2z}\right)^n + 1 - \sum_{k=0}^{\infty} z^k$$

Both the sums in this example must converge is order for X(z) to converge. This implies that we must have

$$\left(\frac{1}{2}\right)\left(\frac{1}{z}\right) < 1 \quad |z| < 1$$

or $|z| > \frac{1}{2}$ &

Hence

$$X(z) \Rightarrow \frac{1}{1-\left(\frac{1}{2}\right) z^{-1}} + 1 - \frac{1}{1-z}\ \frac{1}{2} < |z| < 1$$

$$= \frac{z\left(2z - \frac{3}{2}\right)}{\left(z-\frac{1}{2}\right)(z-1)}\ \frac{1}{2} < |z| < 1$$

ROC is a ring in the z-plane.

Example 5:

$$x[n] \left(\frac{1}{2}\right)^n u[n] + \left(\frac{-1}{3}\right)^n u[n]$$

Answer is $X(z) = \dfrac{z\left(2z - \frac{1}{6}\right)}{\left(z - \frac{1}{2}\right)\left(z + \frac{1}{3}\right)}$

$|z| > \frac{1}{2}$

Solve at hence.

Properties of Region of Convergence

In this, we show how the ROC is related to the characteristics of a signal, x[n]. The ROC properties, we can often identify the ROC from X(z) and limited knowledge of the characteristics of x[n]. The relationship between the ROC and the characteristics of the fine domain signal is used to find the increase z-transform.

First we know that the ROC cannot can train only poles. This is because ROC is defined as the set of z for which the z-transform converges. Hence X(z) must be finite for all z in the ROC. However if d is a pole, then $|x(\alpha)| = \infty$ and the z-transform does not converge at the pole. Thus the pole cannot lie is the ROC.

Second : The ROC for a finite duration signal includes the entire z-plane, except possibly z = 0 and/or $|z| = \infty$. To see this, suppose x[n] is non-zero only on the interval $n_1 \le n \le n_2$, We have

$$X(z) = \sum_{n = n_1}^{n_2} x[n]\, z^{-n}$$

This sum will converge provided that each term in the sum in finite.

Now consider infinite-duration signals. The condition for convergence is |X(z)| < ¥ we may thus write

$$|X(z)| = \left| \sum_{n = n-\infty}^{\infty} x[n]\, z^{-n} \right]$$

$$\le \sum_{n = n-\infty}^{\infty} \left|x[n]\, z^{-n}\right|$$

$$= \sum_{n = n-\infty}^{\infty} |x[n]|\, \left|z^{-n}\right|$$

The second line follows from the fact that the magnitude of a sum of numbers is less than or equal to sum of magnitudes.

We obtain the third line by noting that the magnitude of a product equal to the product of magnitudes. Splitting the unifinite sum into negative and positive time portions. We may write

$$X_{-}(z) = \sum_{n = n-\infty}^{-1} |x(n)| \, |z^{-n}|$$

and $$X_{+}(z) = \sum_{n=0}^{\infty} |x(n)| \, |z^{-n}|$$

We note that $|X(z)| \leq X_{-}(z) + X_{+}(z)$. If both $X_{-}(z)$ and $X_{+}(z)$ are finite, then $|X(z)|$ is guaranteed finite. This clearly shows that $|x[n]|$ be bounded is some way. Suppose we can bound $|x[n]|$ by finding small positive constants A_{-}, A_{+}, r_{-}, r_{+} such that

$$|x[n]| \leq A_{-}(r_{-})^n \quad n < 0 \qquad \text{...(1)}$$

and $$|x[n]| \leq A_{+}(r_{+})^n \quad n \geq 0 \qquad \text{...(2)}$$

A signal that satisfies these bounds grows no faster than $(r_{+})^n$ for positive n and $(r_{-})^n$ for negative n. While we can construct signals that do not satisfy these bounds, such as a^{n^2}, such type of signal do not appear in an engineering problems.

If bound is satisfied then

$$X_{-}(z) \leq A_{-} \sum_{n=-\infty}^{\infty} (r_{-})^n \, |z^{-n}|$$

$$= A_{-} \sum_{n=-\infty}^{-1} \left(\frac{r_{-}}{|z|}\right)^n$$

$$= A_{-} \sum_{n=1}^{\infty} \left(\frac{|z|}{r_{-}}\right)^k$$

This last sum converges if and only if

$|z| < r_{-}$. Now consider positive time function

$$X_{+}(z) \leq A_{+} \sum_{n=0}^{\infty} (r_{+})^n \, |z|^{-n}$$

$$A_{+} \sum_{n=0}^{\infty} \left(\frac{r_{+}}{|z|}\right)^n$$

This sum converges if and only if $|z| > r_{+}$. Hence if $r_{+} < |z| < r_{-}$ then both $X_{+}(z)$ and $X_{-}(z)$ converges and $|X(z)|$ also converges. Note that if $r_{+} > r_{-}$ then there are no values of z for which convergence is guaranteed.

This analyses supports the following conclusions for infinite duration signals. Define self-sided signal as one for which x[n] = 0 for $n > n_2$ and right sided signal as one for which x[n] = 0 for n < n, and two sided signal a sone that has finite value (duration) is both positive and negative directions. For signals x[n] that satisfy the experiential bounds in equation 1 & 2 we have

1. The ROC of a right sided signal is of the form $|z| > r_+$.
2. The ROC of a left sided signal is of the form $|z| < r_-$.
3. The ROC of a two sided signal is of the from

   $r_+ < |z| < r_-$

One sided experiential signals are frequently encountered in engineering problem. In this case the magnitude of one or more poles determines the ROC boundaries r_- and r_+.

Suppose we have a right handed signal $x[n] = \alpha^n u[n]$ The z-transform of x[n] has a pole at $z = \alpha$ and the ROC is $|z| > |\alpha|$ Hence the ROC is the region of the z plane with a radius greater than the radius of pole. Like wise if $x[n] = \alpha^n u[-n-1]$ (a left sides f^n), the ROC is $|z| < |\alpha|$; the region of the z-plane with radius less than the radius of the pole.

Example 6:

(i) $z\left(na^n\right) = \dfrac{az}{(z-a)^2}$

(ii) $z\left(n^2a^n\right) = \dfrac{az^2 + a^2z}{(z-a)^3}$

(iii) $z(\cos n\theta) = \dfrac{z(z-\cos\theta)}{z^2 - 2z\cos\theta + 1}$

(iv) $z(\sin n\theta) = \dfrac{z\sin\theta}{z^2 - 2z\cos\theta + 1}$

(v) $z(a^n \cos n\theta) = \dfrac{z(z - a\cos\theta)}{z^2 - 2az\cos\theta + a^2}$

(vi) $z(a^n \sin n\theta) = \dfrac{az\sin\theta}{z^2 - 2az\cos\theta + a^2}$

Solution:

(i) We know that $z(n) = \dfrac{z}{(z-1)^2}$

$$z(na^n) = \bar{u}\left(a^{-1}z\right) \Rightarrow \frac{a^{-1}z}{\left(a^{-1}z-1\right)^2}$$

$$\Rightarrow \frac{az}{(z-a)^2}$$

$$\text{(ii) } z(n^2) = \frac{z^2+z}{(z-1)^3} = \bar{u}(z)$$

$$z\,(n^2a^n) = \bar{u}\,(a^{-1}z) \Rightarrow \frac{\left(a^{-1}z\right)^2 + a^{-1}z}{\left(a^{-1}z-1\right)^3}$$

$$= \frac{\left(a^{-1}z\right)^2 + a^{-1}z}{\left(a^{-1}z-1\right)}.$$

We know that $z\,(\cos n\theta)$

$$= \frac{z(z-\cos\theta)}{z^2 - az\cos\theta + 1}$$

$$z(a^n \cos n\theta) = \frac{a^{-1}z\left(a^{-1}z - \cos\theta\right)}{\left(a^{-1}z\right)^2 - z\left(a^{-1}z\right)\cos\theta + 1}$$

$$= \frac{z(z-a\cos\theta)}{z^2 - 2az\cos\theta + a^2}.$$

(i) Shifting u_n to the right

If $z(u_n) = \bar{u}(z)$ then $z(u_{n-k}) = z^{-k}\,\bar{u}(z)$ for $k > 0$

By definition

$$z(u_{n-k}) = \sum_{n=0}^{\infty} u_{n-k}\, z^{-n}$$

$$= u_0\, z^{-k} + u_1 z^{-(k+1)} + \ldots$$

$$= z^{-k} \sum_{n=0}^{\infty} u_n\, z^{-n} = z^{-k}\,\bar{u}(z)$$

This rule is applicable/or very useful to difference equations.

(ii) Shifting u_n to the left

of $z(u_n) = \bar{u}(z)$ then

$$z(u_{n+k}) = z^k\, [\overline{u}(z) - u_0 - u_1 z^{-1} - u_2 z^{-2} \ldots - u_{k-1}\, z^{-)k-1)}].$$

Solution:

$$z(u_{n+k}) = \sum_{n=0}^{\infty} u_{n+k}\, z^{-n}$$

$$= z^k \sum_{n=0}^{\infty} u_{n+k}\, z^{-n} = z^{-(n+k)}$$

$$= z^k \left[\sum_{n=0}^{\infty} u_n\, z^{-n} - \sum_{n=0}^{k-1} u_n\, z^{-n} \right]$$

Hence $z(u_{n+k}) = z^k\, \overline{u}(z) - u_0 - u_1 z^{-1} - u_2 z^{-2} \ldots - u_{k-1}\, z^{-(k-1)}]$

The following useful results are

(i) $z(u_{n+1}) = z(\overline{u}(z) - u_0)$

(ii) $z(u_{n+k}) = z^2(\overline{u}(z) - u_0 - u_1 z^{-1})$

(iii) $z(u_{n+3}) = z^3(\overline{u}(z) - u_0 - u_1 z^{-1} - u_2 z^{-2})$.